Withdrawn from
Davidson College Library

Library of
Davidson College

OCCASIONAL PAPER
64

The Federal Republic of Germany
Adjustment in a Surplus Country

By Leslie Lipschitz, Jeroen Kremers, Thomas Mayer, and Donogh McDonald

International Monetary Fund
Washington, D.C.
January 1989

Recent Occasional Papers of the International Monetary Fund

42. Global Effects of Fund-Supported Adjustment Programs, by Morris Goldstein. 1986.
43. International Capital Markets: Developments and Prospects, by Maxwell Watson, David Mathieson, Russell Kincaid, and Eliot Kalter. 1986
44. A Review of the Fiscal Impulse Measure, by Peter S. Heller, Richard D. Haas, and Ahsan H. Mansur. 1986.
45. Switzerland's Role as an International Financial Center, by Benedicte Vibe Christensen. 1986.
46. Fund-Supported Programs, Fiscal Policy, and Income Distribution: A Study by the Fiscal Affairs Department of the International Monetary Fund. 1986.
47. Aging and Social Expenditure in the Major Industrial Countries, 1980–2025, by Peter S. Heller, Richard Hemming, Peter W. Kohnert, and a Staff Team from the Fiscal Affairs Department. 1986.
48. The European Monetary System: Recent Developments, by Horst Ungerer, Owen Evans, Thomas Mayer, and Philip Young. 1986.
49. Islamic Banking, by Zubair Iqbal and Abbas Mirakhor. 1987.
50. Strengthening the International Monetary System: Exchange Rates, Surveillance, and Objective Indicators, by Andrew Crockett and Morris Goldstein. 1987.
51. The Role of the SDR in the International Monetary System, by the Research and Treasurer's Departments of the International Monetary Fund. 1987.
52. Structural Reform, Stabilization, and Growth in Turkey, by George Kopits. 1987.
53. Floating Exchange Rates in Developing Countries: Experience with Auction and Interbank Markets, by Peter J. Quirk, Benedicte Vibe Christensen, Kyung-Mo Huh, and Toshihiko Sasaki. 1987.
54. Protection and Liberalization: A Review of Analytical Issues, by W. Max Corden. 1987.
55. Theoretical Aspects of the Design of Fund-Supported Adjustment Programs: A Study by the Research Department of the International Monetary Fund. 1987.
56. Privatization and Public Enterprises, by Richard Hemming and Ali M. Mansoor. 1988.
57. The Search for Efficiency in the Adjustment Process: Spain in the 1980s, by Augusto Lopez-Claros. 1988.
58. The Implications of Fund-Supported Adjustment Programs for Poverty: Experiences in Selected Countries, by Peter S. Heller, A. Lans Bovenberg, Thanos Catsambas, Ke-Young Chu, and Parthasarathi Shome. 1988.
59. The Measurement of Fiscal Impact: Methodological Issues, edited by Mario I. Blejer and Ke-Young Chu. 1988.
60. Policies for Developing Forward Foreign Exchange Markets, by Peter J. Quirk, Graham Hacche, Viktor Schoofs, and Lothar Weniger. 1988.
61. Policy Coordination in the European Monetary System. Part I: The European Monetary System: A Balance Between Rules and Discretion, by Manuel Guitián. Part II: Monetary Coordination Within the European Monetary System: Is There a Rule? by Massimo Russo and Giuseppe Tullio. 1988.
62. The Common Agricultural Policy of the European Community: Principles and Consequences, by Julius Rosenblatt, Thomas Mayer, Kasper Bartholdy, Dimitrios Demekas, Sanjeev Gupta, and Leslie Lipschitz. 1988.
63. Issues and Developments in International Trade Policy, by Margaret Kelly, Naheed Kirmani, Miranda Xafa, Clemens Boonekamp, and Peter Winglee. 1988.
64. The Federal Republic of Germany: Adjustment in a Surplus Country, by Leslie Lipschitz, Jeroen Kremers, Thomas Mayer, and Donogh McDonald. 1989.

Note: For information on the titles and availability of Occasional Papers published prior to 1986, please consult the most recent IMF *Publications Catalog* or contact IMF Publication Services.

Occasional Paper No. 64

The Federal Republic of Germany
Adjustment in a Surplus Country

By Leslie Lipschitz, Jeroen Kremers,
Thomas Mayer, and Donogh McDonald

International Monetary Fund
Washington, D.C.
January 1989

© 1989 International Monetary Fund

Library of Congress Cataloging-in-Publication Data

The Federal Republic of Germany : adjustment in a surplus country / by Leslie Lipschitz . . . [et al.].
    p. cm. — (Occasional paper, ISSN 0251-6365 ; no. 64)
    "January 1989."
    Bibliography: p.
    ISBN 1-55775-088-2
    1. Germany (West)—Economic conditions—1974- 2. Germany (West)—Economic policy—1974- 3. Economic forecasting—Germany (West) I. Lipschitz, Leslie. II. Series: Occasional paper (International Monetary Fund) ; no. 64.
HC286.7.F375   1989
338.943—dc19   89-1896
                                                                          CIP

Price: US$7.50
(US$4.50 university libraries, faculty members, and students)

Address orders to:
External Relations Department, Publication Services
International Monetary Fund, Washington, D.C. 20431

Contents

		Page
Preface		**ix**
I.	**Introduction**	**1**
II.	**Recent Economic Developments, Prospects, and Policy Issues**	**2**

      Recent Economic Developments
         Domestic Demand
         The Foreign Balance
         Production, Employment, and Prices
      Economic Prospects and Policy Issues

III.	**Demand Management Policies and Financial Market Developments**	**8**

      Monetary Policy and Financial Market Developments
         Monetary Developments, 1986–88
         Interest Rate Developments
         Capital Market Developments and the Capital Account of
           the Balance of Payments
         The Future Development of the EMS
      Fiscal Policy
         Principles
         Achievements
         The Outlook
         Evaluation
      Policy Issues

IV.	**Structural Policies**	**29**

      The Slowdown in Economic Growth
         Labor
         Capital
         Total Factor Productivity
      Tax Reform
      Reform of the Social Security System
         The Pension System
         Medical Insurance
         Unemployment Insurance
      Trade and Industrial Policies
         Agriculture
         Coal Mining
         Steel
         Shipbuilding
         Other Industries

CONTENTS

Page

      Privatization and Deregulation
         Financial Services
         Telecommunications
         Transportation
         Retail Trade
         Professional Services
      Economic Outlook Under Alternative Policy Scenarios
         The Methodology
         Simulations

Appendices

I.	**Statistical Tables**	55
II.	**Federal Republic of Germany: A Dynamic Macro-Model**	96
III.	**Federal Republic of Germany: A Computable General Equilibrium Model**	98
Bibliography		102

TABLES
Section

II.	1. The Foreign Balance and the Terms of Trade	3
	2. Merchandise Export Indicators	5
	3. Merchandise Import Indicators	5
	4. Economic Prospects	6
III.	5. Medium-Term Fiscal Projections for the General Government	26
	6. Fiscal Balances Excluding Profit Transfers from the Bundesbank	26
	7. Structural Budget Balances for the Federal and General Governments	27
IV.	8. Direct and Indirect Taxes in Percent of Total	38
	9. Financial Implications of the Health System Reform	40

Appendix

I.	A1. Federal Republic of Germany: Aggregate Demand	55
	A2. Federal Republic of Germany: Labor Market Developments	56
	A3. Federal Republic of Germany: National Income	57
	A4. Federal Republic of Germany: Earnings, Wages, Employment, and Productivity	57
	A5. Federal Republic of Germany: Income and Consumption	58
	A6. Federal Republic of Germany: Balance of Payments Summary	58
	A7. Federal Republic of Germany: Merchandise Exports by Region	59
	A8. Federal Republic of Germany: Merchandise Exports by Product Category	59
	A9. Federal Republic of Germany: Merchandise Imports by Region	60

Page

A10.	Federal Republic of Germany: Merchandise Imports by Product Category	60
A11.	Federal Republic of Germany: Regional Breakdown of Foreign Trade	61
A12.	Federal Republic of Germany: Service Account Receipts and Payments	62
A13.	Federal Republic of Germany: Real Value Added by Sector	62
A14.	Federal Republic of Germany: Value-Added Deflators by Sector	63
A15.	Federal Republic of Germany: Growth of the Main Monetary Aggregates	64
A16.	Federal Republic of Germany: Indicators of the Internationalization of the German Capital Markets	65
A17.	Federal Republic of Germany: Exchange Rates	66
A18.	Federal Republic of Germany: Changes in Central Bank Money and Its Determinants	67
A19.	Federal Republic of Germany: Interest Rates	68
A20.	Federal Republic of Germany: Sales and Purchases of Securities	69
A21.	Federal Republic of Germany: Long-Term Capital in the Balance of Payments Accounts	70
A22.	Federal Republic of Germany: Short-Term Capital in the Balance of Payments Accounts	71
A23.	Federal Republic of Germany: Distribution of the Ownership of Shares, December 1987	71
A24.	Federal Republic of Germany: Selected Indicators of the Size of the General Government	72
A25.	Federal Republic of Germany: Territorial Authorities' Finances	73
A26.	Federal Republic of Germany: Federal Government Budget	74
A27.	Federal Republic of Germany: Länder Government Finances	75
A28.	Federal Republic of Germany: Municipalities' Finances	75
A29.	Federal Republic of Germany: Territorial Authorities—Tax Revenue	76
A30.	Federal Republic of Germany: General Government Revenue and Expenditure	77
A31.	Federal Republic of Germany: Outstanding Debt of the Territorial Authorities	78
A32.	General Government Current Receipts of Major Industrial Countries	79
A33.	Federal Republic of Germany: The 1990 Tax Cuts	80
A34.	Federal Republic of Germany: The 1990 Cuts in Tax Exemptions	81
A35.	International Comparison of Recent and Proposed Changes in Personal Income Tax Systems	82
A36.	International Comparison of Social Security Transfers as a Percentage of GDP	83
A37.	Dependency Ratios in Seven Major Countries	84
A38.	Federal Republic of Germany: Expenditures of the	

CONTENTS

Page

	Medical Insurance System and Development of Macroeconomic Aggregates	84
A39.	Federal Republic of Germany: Revenues and Expenditures of the Medical Insurance System	84
A40.	Federal Republic of Germany: Nominal and Effective Protection in 1982	85
A41.	Federal Republic of Germany: Nontariff Barriers to Trade in 1982	86
A42.	Import Restrictions in the EC in 1982–83 Under Article 115 of the EEC Treaty	86
A43.	International Comparison of Subsidies	86
A44.	Federal Republic of Germany: Subsidies	87
A45.	Federal Republic of Germany: Fiscal Assistance and Tax Relief of the Federal Government	88
A46.	Federal Republic of Germany: Subsidies Provided by the Federal Government for Selected Industries	89
A47.	Key Features of the Agricultural Sector in the EC in 1986	89
A48.	Final Agricultural Production, Consumption of Inputs, and Gross Value Added in the EC	90
A49.	Employment by Sector in the EC	90
A50.	Net Agricultural Value Added in the EC at Factor Cost per Manpower Unit	90
A51.	Federal Republic of Germany: Coefficients of Nominal Protection for Selected Agricultural Products	91
A52.	Federal Republic of Germany: Degrees of Self-Sufficiency in Selected Agricultural Products	91
A53.	EC Common Agricultural Policy: Budgetary Expenditures of the EAGGF-Guarantee Fund by Commodities	91
A54.	Federal Republic of Germany: Payments to and Receipts from the EC	92
A55.	Intervention Prices in the EC in 1987/88 for Member Countries	92
A56.	Federal Republic of Germany: Real Output and Input Prices in Agriculture	92
A57.	Rates of Capacity Utilization in the EC Steel Industry, 1970–83	93
A58.	Employment in the EC Steel Industry	93
A59.	Subsidies to the EC Steel Industry, 1980–85	93
A60.	Finished Steel Products in the EC: Quotas and Steel Production	93
A61.	Shipbuilding Capacity in the EC, 1976–86	93
A62.	Employment in the EC Shipbuilding Industry, 1975–86	94
A63.	Real Stock Market Capitalization in Selected Industrial Countries	94
A64.	Federal Republic of Germany: Projections for 1990–91 Under Alternative Scenarios—Macroeconomic Variables	94
A65.	Federal Republic of Germany: Projections for 1991 Under Alternative Scenarios—Selected Sectoral Variables	95

Contents

Page

II.	A66. Federal Republic of Germany: A Dynamic Macro-Model	96
	A67. Federal Republic of Germany: Exogenous Variables of the Dynamic Macro-Model	97
III.	A68. Federal Republic of Germany: A CGE Model	98
	A69. Federal Republic of Germany: Notation of the Variables of the CGE Model	100
	A70. Federal Republic of Germany: Selection and Values of the Exogenous Variables of the CGE Model	101

CHARTS
Section

II.	1. Real GNP and Real GNP Per Capita in the Federal Republic of Germany and Selected Industrial Countries	2
	2. Federal Republic of Germany: Exchange Rate Developments	2
	3. National Saving in the Federal Republic of Germany and Selected Industrial Countries	3
	4. Federal Republic of Germany: Saving-Investment, Current Account, and Fiscal Balances	4
	5. Federal Republic of Germany: Employment and Unemployment	6
III.	6. Federal Republic of Germany: Central Bank Money and Broad Money (M3)	9
	7. Federal Republic of Germany: Interest Rate Developments	14
	8. Federal Republic of Germany: Interest Rates in Historical Perspective	15
	9. Federal Republic of Germany: Short-Term U.S. Dollar and Deutsche Mark Interest Rates	16
	10. Federal Republic of Germany: Selected Indicators of Fiscal Policy	21
	11. Federal Republic of Germany: Contribution of Public Consumption and Investment to Growth of Real GNP	21
IV.	12. Federal Republic of Germany: The Beveridge Curve	31
	13. Federal Republic of Germany: The Okun Curve	31
	14. Federal Republic of Germany: Capital/Labor Ratios	32
	15. Federal Republic of Germany: The Wedge Between the Net Consumption Wage and the Gross Product Wage	33
	16. Federal Republic of Germany: Nominal Gross Fixed Investment, 1960–87	34
	17. Federal Republic of Germany: The Components of Real Private Gross Fixed Investment, 1960–87	34
	18. Federal Republic of Germany: Growth of Real Business Sector Net Capital Stock and Its Components, 1960–87	34
	19. Federal Republic of Germany: Capital Intensity of Production, Relative Factor Prices, and the Return to Capital	35

CONTENTS

Page

20. Federal Republic of Germany: Tax Reform and Marginal Income Tax Rates 37
21. Federal Republic of Germany: Yields on Deutsche Mark Bonds Issued by Domestic and Foreign Public Authorities 48

The following symbols have been used throughout this paper:
... to indicate that data are not available;
— to indicate that the figure is zero or less than half the final digit shown, or that the item does not exist;
– between years or months (e.g., 1984–85 or January–June) to indicate the years or months covered, including the beginning and ending years or months;
/ between years (e.g., 1985/86) to indicate a crop or fiscal (financial) year.
"Billion" means a thousand million.
Minor discrepancies between constituent figures and totals are due to rounding.

Preface

This study reviews and analyzes economic developments and policies in the Federal Republic of Germany during the 1980s. It traces the dynamics of the cyclical upswing that began at the end of 1982 and describes the genesis of the two major imbalances in the German economy in the course of the 1980s: high unemployment and large external surpluses. A discussion of the economic policies adopted to deal with these imbalances forms the main body of the paper. The analysis of these policies highlights the constraints on adjustment for a country with external surpluses. The study then concludes with an illustrative exercise on the medium-term economic outlook for the German economy under alternative policy scenarios. The results of this exercise underline the importance of structural policy measures in dealing with the disequilibria in the German economy.

The study was inspired by the 1988 Article IV consultation discussions of the Fund staff with the German authorities (concluded in July of that year). The opinions expressed, however, are those of the authors alone and should not be construed as reflecting the views of the German authorities, Executive Directors of the Fund, or other staff members. The authors also bear sole responsibility for any factual errors. The inevitable time lag between preparation and publication unfortunately means that some of the data will be out of date by the time of publication—the statistical material contained in Appendix I reflects information available in the summer of 1988. Similarly, information that became available during the autumn of 1988 could only be partially incorporated into the text of the paper. Recent revisions of the data and estimates, however, do not materially alter the broad conclusions of the paper.

The authors are indebted to Manuel Guitián for his support in the preparation of the study and many helpful suggestions and comments, to Aurel Schubert for helpful background work on structural issues in Germany, to Behrouz Guerami for research assistance, to Juanita Roushdy of the External Relations Department for editorial assistance, and to Hilda Newman, Valerie Pabst, and Ann Phillips for secretarial support.

I Introduction

The current expansion in the Federal Republic of Germany, which began at the end of 1982, has now completed its sixth year. Late phases of cyclical expansions are often characterized by a variety of stresses and imbalances—production bottlenecks, rising costs and prices, falling profits, lower savings ratios, and higher costs of financing new investment—that presage a downturn, but so far no appreciable negative signals of this sort are evident in the German economy. Therefore, unless there are disruptions from abroad, developments appear to be on track for a continued, if unspectacular, expansion of economic activity, in an environment of low inflation.

The recovery has been characterized, however, by large imbalances in the labor market and in the external accounts. The unemployment rate has risen from 1 percent in 1970 to 3¼ percent in 1980 and to 7¾ percent in 1988, and no immediate prospect of a substantial reduction in unemployment is in sight. The other imbalance, the current account surplus, is similarly resilient. The surplus rose from about ¾ of 1 percent of gross national product (GNP) in 1982 to 4½ percent in 1986; it is estimated at about 4 percent of GNP in 1988 and it seems likely that large surpluses will persist in the near future. At first glance, the current account surplus appears to be more a problem from an international perspective than from the perspective of a German citizen; however, sizable international current account imbalances are sustainable only as long as they can be financed, and a prolonged period of such large imbalances must raise the probability of instability in global financial markets that could be disruptive to continued economic growth in Germany, as well as in the world as a whole.

Section II contains a review of recent economic developments and economic prospects in which the emergence of the principal imbalances is traced and their persistent character highlighted. The remaining sections focus on economic policy with a view to answering the questions: What could policies have done to foster more balanced growth? and What ought they to do now? Section III analyzes macroeconomic policies—the intellectual framework in which they are set, their recent history, and the limits to their efficacy. Section IV examines the growth performance of the German economy from the supply side with particular emphasis on microeconomic policies and structural rigidities. Here, it is necessary to discuss, in some detail, the laws, regulations, and practices that govern economic activity in a wide variety of sectors. To avoid losing the reader in the detail and in order to reach a judgment on the costs and benefits of the various microeconomic characteristics touched upon, a quantitative model is used to analyze the effects of alternative microeconomic structures.

On macroeconomic policies, while one might quibble about the rate of expansion of the money supply or the timing of tax reduction and reform, the paper does not find major errors in the formulation or implementation of policies that would account for the principal economic imbalances. The conclusions on microeconomics are much stronger. Myriad rigidities and disincentives affecting the labor market, agriculture, mining, shipbuilding, iron and steel, textiles, commercial aircraft, insurance, retail trade, telecommunications, and professional services, coupled with government policies on taxes, subsidies and trade restrictions, play a significant role in stifling the growth of output, investment, and employment, and reducing the responsiveness of the economy to market signals.

This is perhaps a rather unhappy conclusion. For, while macroeconomic policies are subject to general scrutiny and debate and errors are thus more likely to be identified and corrected, microeconomic problems are buried in sectoral and industry details and are often lodged in institutional practices. Nevertheless, for a government committed to establishing a sound but unobtrusive framework within which the economy can respond readily to market forces, the findings of this paper do suggest a full agenda for reform, albeit a difficult and protracted one.

II Recent Economic Developments, Prospects, and Policy Issues

Recent Economic Developments

Between 1982 and 1987 real GNP in the Federal Republic of Germany grew at an average annual rate of 2¼ percent (Table A1). The rate of growth of GNP per capita was not out of line with that in the other major European countries and the United States (Chart 1) but lower than in previous upswings. Notwithstanding the expansion of economic activity, unemployment, which during the preceding recession had more than doubled from 3¼ percent of the labor force in 1980 to 8 percent in early 1983, remained at this high level (Table A2).

After investment expenditure provided the initial impulse to growth, the external sector was the principal driving force in the economy in 1984 and 1985; strong external demand and a real depreciation of the deutsche mark stimulated output (Chart 2). In 1986 and 1987,

Chart 1. Real GNP and Real GNP Per Capita in the Federal Republic of Germany and Selected Industrial Countries

(1980 = 100)

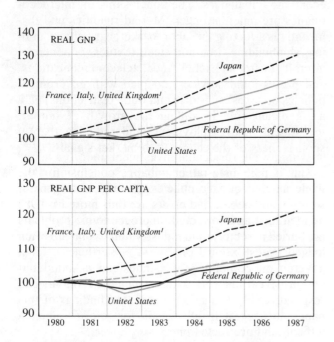

Sources: International Monetary Fund, Data Fund; and Fund staff calculations.
[1] Unweighted average

Chart 2. Federal Republic of Germany: Exchange Rate Developments

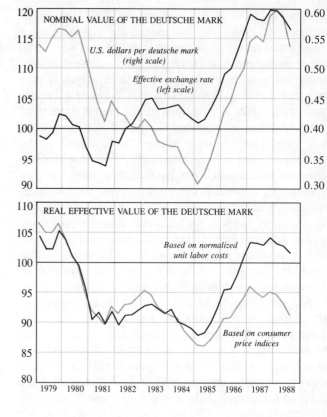

Source: International Monetary Fund, Data Fund.

Table 1. The Foreign Balance and the Terms of Trade[1]

(In percent of GNP)

	1985	1986	1987
Change in nominal foreign balance	1.4	1.7	−0.1
Terms of trade gain[2]	0.3	2.7	0.9
Volume change[3]	1.1	−1.0	−1.0

Source: Statistisches Bundesamt, *Volkswirtschaftliche Gesamtrechnungen*, various issues.
[1] On a national accounts basis (including goods and services).
[2] Based on quantities of exports and imports in the previous year.
[3] Valued at prices of exports and imports in the previous year.

domestic demand, supported by a tax cut and a large terms of trade gain, became the engine of growth, while there was a significant withdrawal of stimulus from the foreign sector. The latter more than offset the increase in the real foreign surplus in the earlier years of the upswing with the result that, between 1982 and 1987, the real external surplus (national accounts basis) declined by ½ of 1 percent of GNP. The nominal current account surplus increased, however, to 4 percent of GNP in 1987, or more than 3 percentage points above its 1982 level, reflecting the large terms of trade gain in 1986–87.

Domestic Demand

The largest component of domestic demand, real private consumption, expanded at a slightly faster pace than real GNP during the course of the upswing. In 1983–85, consumption grew at a modest annual average rate of 1½ percent, reflecting the relatively slow growth of earnings and of household disposable income (Tables A3–A5). In 1986–87, however, the growth of private consumption accelerated to an average rate of 3½ percent, supported by an income tax cut in 1986 (½ of 1 percent of GNP) and a terms of trade gain (equivalent to 3½ percent of GNP over the two years—Table 1).[1] In contrast to private consumption, public sector consumption grew more slowly than GNP during the upswing, reflecting the program of expenditure restraint instituted by the government that took office in October 1982.[2]

Between 1982 and 1985, a rise in the national saving rate reflected the budget consolidation process. In the period 1986–87, private consumption rose more rapidly than GNP; nevertheless, as the increase in the con-

Chart 3. National Saving in the Federal Republic of Germany and Selected Industrial Countries

(Percent of national disposable income)[1]

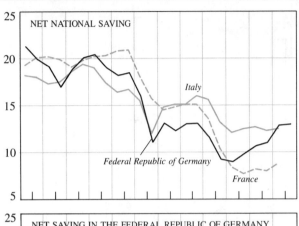

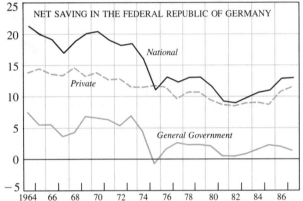

Sources: Organization for Economic Cooperation and Development, *National Accounts;* various issues; and Statistisches Bundesamt, *Volkswirtschaftliche Gesamtrechnungen,* various issues.
[1] National Disposable Income equals Net National Saving plus National Final Consumption.

sumption price deflator was much smaller than that in the GNP deflator (because of the terms of trade gain), a significant further rise in the net national saving rate occurred (Chart 3).[3]

Despite the sharp increase in investable resources, fixed investment declined relative to GNP during the upswing; weak residential investment and, to a lesser extent, public investment were responsible for the decrease. Indeed, business fixed investment grew at an average annual rate of 3¾ percent between 1982 and 1987, much above the rate of growth of GNP. The time pattern of business investment during the upswing was distorted by special investment incentives that affected the recorded growth rates in 1983–85.[4] It is

---

[1] The terms of trade gain is calculated as the difference between the external surplus evaluated at actual export and import prices and the surplus that would have resulted if there had been no change in export and import prices.

[2] In fact, public investment and transfers were more severely affected by expenditure restraint than public consumption (see Section III).

[3] A small decline in the share of nominal depreciation in GNP also contributed to the rise in the net national saving rate over 1982–87.

[4] A special scheme encouraged businesses to bring their investment in machinery and equipment forward from 1984 to 1983 and their construction investment from 1985 to 1984. The actual rates of growth of private investment in machinery and equipment in 1983,

Chart 4. Federal Republic of Germany: Saving-Investment, Current Account, and Fiscal Balances
(In percent of GNP)

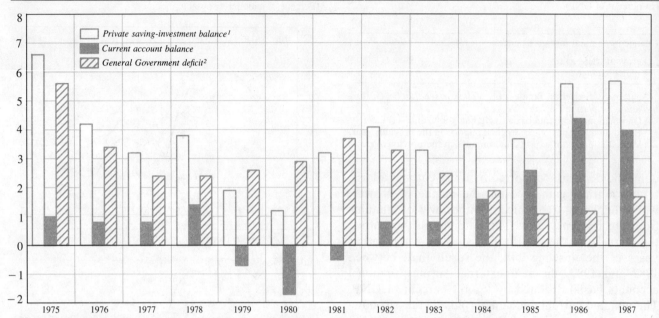

Sources: Statistisches Bundesamt, *Volkswirtschaftliche Gesamtrechnungen* and *Wirtschaft und Statistik,* various issues.
[1] Private saving minus gross private investment, which equals the sum of the general government deficit and the current account balance.
[2] National accounts basis.

notable, however, that, in 1986–87, businesses responded to the sharp increase in investable resources (owing chiefly to the favorable shift in the terms of trade) by building up their liquid financial balances rather than by accelerating fixed investment programs. (Investment behavior is examined in a longer perspective in Section IV.)

The Foreign Balance

Higher national saving relative to GNP coupled with a lower fixed investment ratio made for a rise in the current account surplus from less than 1 percent of GNP in 1982 to 4 percent in 1987 (Chart 4 and Table A6).[5] However, this increase in the current account surplus reflected entirely the terms of trade gain in 1986–87; the foreign surplus, expressed in 1980 prices, fell by ½ of 1 percentage point of GNP between 1982 and 1987 (national accounts basis).[6] Although, for the upswing as a whole, the foreign sector exerted a small negative influence on growth, it was nonetheless an important driving force in the early part of the expansion. In both 1984 and 1985 the contribution of the foreign sector to real GNP growth amounted to about 1 percentage point, while, in each of the two subsequent years, movements in the foreign balance reduced real growth by about 1 percentage point.

In 1984–85, the volume of merchandise exports grew at an annual average rate of 7½ percent (customs basis),[7] stimulated by strong growth of partner-country demand and the improvement of Germany's competitiveness owing to the depreciation of the deutsche mark (Table 2). Over the same period, merchandise imports grew at an annual average rate of 4¾ percent (Table 3), reflecting buoyant demand for intermediate and investment goods, which was related in part to the strength of exports.[8]

In 1986–87, the real growth of merchandise exports slowed (to an annual rate of 2 percent) in large part in reaction to the appreciation of the deutsche mark; in

---

1984, and 1985 were 6, –½ of 1, and 9 percent, respectively. Calculations in McDonald (1988a) suggest that, in the absence of the special investment scheme, growth rates would have been on the order of 2½, 6, and 5½ percent in these three years, respectively. An incentive scheme also affected the timing of residential construction.

[5] Not all of the increased surplus of national saving over fixed investment went into the current account; there was also a turnaround in stockbuilding from –½ of 1 percent of GNP in 1982 to ½ of 1 percent of GNP in 1987.

[6] The foreign balance on a national accounts basis excludes transfers, while the data on the current account, from the external accounts, include transfers. Nominal transfers abroad declined by about ¼ of 1 percent of GNP over this period.

[7] There are various discrepancies in trade volumes between customs data and national accounts data; for example, the customs data exclude trade with the German Democratic Republic and use unit values, while the national accounts include trade with the German Democratic Republic and use base-period-weighted price indices.

[8] Detailed regional and product breakdowns of exports, imports, and trade balances are given in Tables A7–A11.

Table 2. Merchandise Export Indicators

(Percent change from previous year)

	1983	1984	1985	1986	1987
Export value	1.1	12.9	10.0	-2.0	0.2
Export volume	-0.3	9.2	5.9	1.3	2.9
Market growth[1]	3.5	9.3	4.7	4.0	5.3
Competitiveness[2]					
Relative export unit value	0.2	-6.8	0.6	10.2	3.6
Real exchange rate[3]	1.5	-2.4	-0.6	9.0	6.8
Profitability of exports[4]	0.2	0.6	0.8	0.8	-0.6
Export unit value	1.4	3.5	3.9	-3.3	-2.7
Unit labor costs[5]	-0.6	1.1	1.8	3.1	2.8
Import unit value	-0.3	5.9	2.6	-16.0	-6.1

Sources: Statistisches Bundesamt, *Aussenhandel*, various issues; Deutsche Bundesbank, *Monthly Report*, various issues; International Monetary Fund, Data Fund.

[1] Non-oil import volume of partner buyers, with partners weighted according to their importance as markets for German exports.
[2] A positive figure indicates a deterioration of Germany's competitiveness.
[3] Period averages based on relative normalized unit labor costs.
[4] Ratio of export prices to producer prices for domestic sales, excluding energy.
[5] In manufacturing.

both of these years, prices of German exports in foreign currency terms rose relative to those of competitor countries, contributing to an erosion of market shares.

Import unit values dropped sharply in 1986–87, reflecting the decline of oil prices in combination with the appreciation of the deutsche mark. In spite of the dramatically improved competitive position of imports in the German market and vigorous domestic demand, the growth rate of import volumes rose by only 1 percentage point in 1986–87; because of the high import content of exports, the slowdown of export growth militated against an acceleration of imports.

The widening of the real external trade surplus was only partially offset by a small terms of trade loss in 1984–85; the nominal trade surplus thus rose steeply in both of these years. With a sharp swing into surplus in the balance on the services account, the current account surplus rose from less than 1 percent of GNP in 1982 to 2½ percent in 1985.

In 1986–87, the contraction of the real trade surplus was more than offset by the very large terms of trade gain, producing a further sharp rise in the nominal trade surplus. Some counterbalance was offered by developments in the services account, which moved from a surplus of DM 5 billion in 1985 to a deficit of DM 7 billion in 1987 (Table A12).[9] Nevertheless, the current account surplus rose to 4½ percent of GNP in 1986, and then narrowed only slightly in 1987.

Production, Employment, and Prices

The appreciation of the deutsche mark since early 1985 in combination with the fall in international commodity prices resulted in more or less stable consumer prices in Germany in 1986 and 1987 (Table A5). The appreciation, however, produced only a moderate shift in resources from the tradables sectors (comprising manufacturing, transport and communication, and agriculture) to the nontradables sectors (services, retail and wholesale trade, and construction).

During 1984 and 1985, gains in price competitiveness and strong growth of partner-country demand boosted the growth of real value added in the tradables sectors to substantially above that of previous years (Table A13). The increase was not at the expense of production in the nontradables sectors: following a modest slowdown in the early 1980s, real value added in services returned to a relatively steady annual rate of growth, somewhat in excess of 3 percent from 1984 to 1987. But, in 1986–87, the appreciation of the deutsche mark that began in 1985 sharply reduced real growth in the sectors producing tradables without eliciting a commensurate expansion of activity in the nontradables sectors.[10]

Relative price developments in 1985–87 did not serve to produce a shift in production from tradables to nontradables; on the contrary, the drop in international commodity prices allowed the value-added deflator for tradables to rise more rapidly than that for nontrad-

Table 3. Merchandise Import Indicators

(Percent change from previous year)

	1983	1984	1985	1986	1987
Import value	3.6	11.3	6.8	-10.8	-1.0
Import unit value[1]	-0.3	5.9	2.6	-16.0	-6.1
Import volume	4.0	5.2	4.2	6.2	5.4
Real GNP	1.9	3.3	2.0	2.5	1.7
Real domestic demand	2.3	2.0	0.9	3.8	2.9
Import competitiveness[2,3]	1.8	-2.9	0.8	19.7	4.6
Real exchange rate[2,4]	1.5	-2.4	-0.6	9.0	6.8

Sources: Deutsche Bundesbank, *Monthly Report, Statistical Supplement 4*, various issues; Statistisches Bundesamt, *Wirtschaft und Statistik*, various issues; and International Monetary Fund, Data Fund.

[1] In deutsche mark terms.
[2] A positive figure indicates that foreign goods became more competitive in Germany.
[3] Ratio of domestic producer prices to import prices.
[4] Based on relative normalized unit labor costs.

---

[9] The turnaround in the balance on services reflected an increased deficit on tourism (owing to higher domestic income and the appreciation of the deutsche mark) and a reduced surplus on investment income (owing chiefly to lower interest rates and the depreciation of receipts denominated in U.S. dollars).

[10] While the data in Table A13 suggest some shift to nontradables production in 1986, as nontradables production grew faster in 1986–87 than in 1985, the pattern is somewhat distorted by investment incentives that shifted construction from 1985 to 1984, thereby depressing the growth rate in 1985 and boosting it in 1986.

ables (Table A14), more than compensating for the higher rate of increase of unit labor costs in the tradables sector. The absence of a decline in the relative price of tradable to nontradable goods in 1986–87, was reflected in an erosion of the international competitiveness of the traded goods sectors in Germany, leading to a slowdown in growth in those sectors as demand shifted to competing products.

Employment and the labor force rose at the same rate during the course of the 1982–87 upswing (annual average of ½ of 1 percent). Thus, having risen from 3¼ percent in 1980 to 8 percent in early 1983, the unemployment rate subsequently remained broadly unchanged (Table A2). The growth of employment was concentrated in the nontradables sectors (exclusively in services); however, not only was the growth modest in relation to the overall unemployment rate, but, viewed in a longer perspective, it was simply a continuation of the upward trend since the 1960s in the share of the labor force employed in the nontradables sectors (Chart 5). There is nothing to suggest that there was a significant redeployment of the labor force in response to real exchange rate movements. By contrast, since the 1960s employment in the sectors exposed to foreign competition has shown a far greater sensitivity to external influences on the German economy. In particular, both of the oil shocks of the 1970s generated marked and lasting reductions in employment in these sectors. On neither occasion did the nontradables sectors contribute noticeably to the absorption of labor released from the production of tradables. (Developments in the labor market are analyzed in greater detail in Section IV.)

Economic Prospects and Policy Issues

Projections prepared by the authors show a continuation of the economic expansion beyond 1987 (Table 4). In 1988, real GNP is estimated to have grown by about 3½ percent—much higher than was expected earlier in the year and up from only 1¾ percent in 1987. Domestic demand is estimated to have risen more rapidly than in 1987. While the diminishing influence of the terms of trade gain exerted a negative impact on aggregate demand growth, it was offset by the impact of the tax cut (0.7 percent of GNP) that took effect in January 1988, favorable winter weather in early 1988, and the stimulus to business investment provided by strong export demand and high capacity utilization. The withdrawal of stimulus from the external sector was significantly smaller than in 1987, reflecting the waning of exchange rate effects on trade volumes and the strength of demand in Germany's trading partners. Despite the higher growth rate, only a small lowering of the unemployment rate occurred—gains in employment were largely matched by growth of the labor force.

Chart 5. Federal Republic of Germany: Employment and Unemployment

(In percent of labor force)

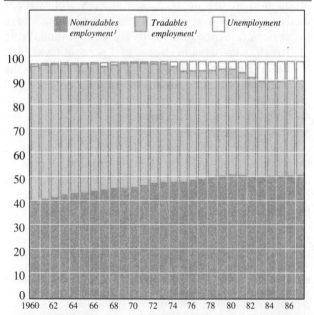

Source: Fund staff calculations based on Statistisches Bundesamt, *Volkswirtschaftliche Gesamtrechnungen,* various issues.
[1] Tradables comprise manufacturing, agriculture, and transport and communication. Nontradables comprise public and private services, construction, and trade. The energy sector is not included.

Table 4. Economic Prospects[1]

(Percentage change)

	1987[2]	1988[3]	1989–90[3] (average)
GNP	1.8	3½	2½
Domestic demand	3.1	3¾	2¾
Final domestic demand	2.8	3¼	2¾
Private consumption	3.5	2½	3
Government consumption	1.6	2	1¼
Gross fixed investment	1.8	6	3¾
Construction	0.2	5¾	2
Machinery and equipment	4.0	6¼	6
Stockbuilding[4]	0.3	½	—
Foreign balance[4]	−1.2	—	—
Exports of goods and services	0.8	5½	4¾
Imports of goods and services	4.9	6½	5¼
Current account (percent of GNP)	4.0	4	3¾[5]
Unemployment (percent of labor force)	7.9	7¾	7½[5]
Consumer price index	0.2	1¼	2½

Source: Statistisches Bundesamt, *Volkswirtschaftliche Gesamtrechnungen* various issues; and Fund staff estimates.
[1] In prices of 1980.
[2] Reflects data revisions not incorporated in the Appendix tables.
[3] Fund staff estimates, December 1988.
[4] Contribution to growth.
[5] In 1990.

Price inflation, while remaining modest, picked up during 1988 to a rate of perhaps 1¾ percent through the year for the consumer price index. This pickup was due chiefly to the dissipation of the moderating terms of trade effect on prices in the previous two years rather than to domestic wage pressures. Nevertheless, inflation should be seen as quiescent rather than vanquished; a combination of higher prices for imports and the rise in excise taxes envisaged in 1989 could disturb the relative peace in the labor market.

With respect to the current account, a weakening of real adjustment was evident, as the influence of past exchange rate changes diminished. Indeed, with demand in Germany's trading partner countries remaining very strong in 1988, exports picked up sharply in volume and provided some of the impetus for more rapid growth. The rate of growth of imports also rose, but this was not sufficient to produce a reduction in the current surplus in proportion to GNP.

Medium-term projections are more tentative. Nevertheless, without a major recession, and assuming growth, on average, at the rate of expansion of potential (about 2½ percent), one would expect some amelioration of the unemployment problem in the early 1990s. Even then, however, the unemployment rate is likely to remain high by historical standards. Moreover, projections on the basis of currently prevailing real exchange rates and continued rapid growth of demand in Germany's major trading partners show further sizable current account surpluses.

The review of recent economic developments and the medium-term projections identify two major imbalances that appear likely to persist for some time: unemployment and external surpluses. The remainder of this study examines the principal areas of economic policy in Germany, with a view to answering the question of how economic policies might be able to redress these imbalances, while, at the same time, promoting durable noninflationary growth.

III Demand Management Policies and Financial Market Developments

Since 1985, macroeconomic policies in Germany have been inextricably bound up with international developments and, in particular, developments in international financial markets. Monetary expansion has exceeded the upper bound of its target range and has been well above the rate of growth of nominal income and, for some of this period, the operating interest rates of the Deutsche Bundesbank were brought down to historically low levels. These developments have reflected, in significant part, a response to both domestic and international concerns about events in foreign exchanges and a desire to reduce the volatility of financial markets. The relationship between international financial developments and the domestic economy is evident in the recent data on German capital markets, capital flows in the balance of payments, and the foreign exchange intervention of the Bundesbank. These interrelationships are discussed in the first part of this section.

Fiscal policy, too, has not been impervious to the international environment. The medium-term fiscal program, initiated by the Government in 1982–83, envisaged a steady lowering of the deficit and the indebtedness of the Government in relation to GNP. More recently, however, concerns about the size of the current account surplus and the strength and sustainability of domestic demand have contributed to a moderation of the effort to consolidate further the public finances. Instead, the Government has moved ahead with its plans to reduce and reform taxation even though this means a temporarily higher public sector deficit and a rise in the ratio of government debt to GNP. The macroeconomic aspects of fiscal policy are discussed later in this section (while the structural, or microeconomic, aspects are taken up in Section IV).

There is no evidence to date that the modification of German economic policies since 1985 has sparked a resurgence of inflationary pressures. Some signs of uncertainty in domestic financial markets have surfaced, however, and, over the medium term, both commitments to the EC and prospective demographic developments argue for a measure of caution in fiscal policy. Against this background, the scope for using aggregate demand policies to reduce unemployment and the external surplus is discussed in the last part of the section.

Monetary Policy and Financial Market Developments

Under Article 3 of the Deutsche Bundesbank Act of 1957, the Bundesbank is required to regulate money and credit with the aim of safeguarding the currency.[11] To this end, the Bundesbank has, from 1975, set monetary targets to conform with the economy's growth potential in conjunction with a rate of inflation that is both achievable and acceptable. The approach is designed to encourage stability and confidence through the transparency, steadiness, and credibility of policy. Moreover, the focus on the growth of potential rather than actual GNP ensures that there is automatic stabilizing feedback from monetary developments to the real economy: when actual growth exceeds the growth of potential, monetary conditions automatically become more restraining, and when actual growth falls short of potential, monetary conditions become more accommodating.[12]

Beginning with 1979, the monetary target has been expressed in the form of a target range for the through-the-year growth (fourth-quarter-to-fourth-quarter) of the targeted aggregate. The midpoint of this target has been calculated as the sum of the growth rate of potential output and a specified acceptable rate of inflation; a margin of error of 1 or 1½ percentage

---

[11] The Bundesbank is also obliged to support the general economic policy of the Federal Government but subject to the provision that this does not prejudice its basic aim of safeguarding the currency (Article 12 of the Bundesbank Act); the Bundesbank is independent of instructions from the Federal Government. See Chapter 1 of Deutsche Bundesbank (1987a) for a discussion of the origins of the Bundesbank, its constitution, and its functions.

[12] Of course, this focus on potential output is not without difficulty. In particular, the concept of productive potential is not an easy one and equally respectable estimates of productive potential may differ.

points has been allowed on either side of the central estimate.[13] For most of the period during which formal monetary targets have been announced, central bank money has been the monetary aggregate targeted by the Bundesbank; for 1988, however, broad money (M3) was selected as the targeted aggregate. Central bank money consists of currency held by nonbanks and the required reserves held by banks against liabilities to residents, calculated on the basis of the minimum reserve ratios in effect in January 1974.[14] Central bank money is thus similar to the monetary base; the statistical difference between the two concepts reflects the fact that the monetary base incorporates required reserves on banks' external liabilities, free reserves held by banks, and any changes in required reserves on domestic liabilities due to changes in required reserve ratios since January 1974. The Bundesbank, however, does not see its system as one of base control. Rather it sees its operations affecting conditions in the money market which subsequently alter banks' liquidity needs and thus the stock of central bank money. From this perspective, since central bank money is essentially a weighted average of the components of M3, the change of targeted aggregate for 1988 was not as large a step as it might appear at first glance.

The formulation of the monetary target underlines the medium-term orientation of monetary policy. While shorter-term macroeconomic considerations are not given prominence in the framing of the target, the Bundesbank does pay attention to factors such as the domestic cyclical position, exchange rate pressures, and money market conditions in implementing monetary policy. In the past, it has usually been possible to accommodate any shorter-term considerations within the target range. Between 1979 (when the monetary target was first expressed in terms of a range) and 1985, monetary expansion never exceeded the upper bound of the target range;[15] in 1980 and 1981, monetary growth fell slightly short of the target, as the deutsche mark weakened considerably against the U.S. dollar. In 1986 and 1987, however, the Bundesbank deviated substantially from its intended course for monetary policy, owing largely to a desire to moderate upward

[13] From 1975 to 1978, the target was a single number; for 1975 the target covered the period December to December and for 1976–78 it was on an annual average basis. For a discussion of the first ten years of monetary targeting, see Deutsche Bundesbank (1985a).

[14] Thus, in calculating central bank money, currency in circulation has a weight of 100 percent, sight deposits have a weight of 16.6 percent, time deposits of less than four years' maturity have a weight of 12.4 percent, and savings deposits of less than four years' maturity, a weight of 8.1 percent.

[15] In 1978, the growth of central bank money exceeded the point target by 3 percentage points, as the Bundesbank was confronted with strong upward pressure on the deutsche mark.

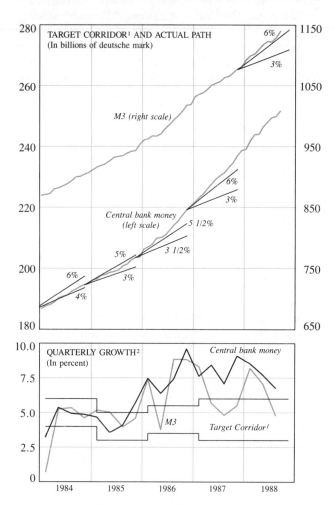

Chart 6. Federal Republic of Germany: Central Bank Money and Broad Money (M3)

Source: Deutsche Bundesbank, *Monthly Report*, Supplement 4, various issues.
[1] The target corridor applied to central bank money up to 1987. For 1988, the corridor applies to M3.
[2] Percent changes from the previous quarter, seasonally adjusted and annualized.

pressure on the deutsche mark;[16] on a fourth-quarter-to-fourth-quarter basis, central bank money grew at a rate of 7¾ percent during 1986 and 8 percent in the course of 1987, compared with target ranges of 3½–5½ percent and 3–6 percent, respectively (Chart 6 and Table A15). Broad money (M3) also grew at rapid rates (7¼ percent during 1986 and 6 percent during 1987), particularly when seen in the context of the much lower rate of growth of nominal GNP. The fast pace of monetary growth continued in the first half of 1988 but some slowing was evident following a tight-

[16] The Bundesbank has indicated that domestic considerations were also a factor. Moreover, in Germany, the exchange rate cannot easily be separated from domestic considerations, as manufacturing is said to be quite sensitive to exchange rate developments and exports have been an important driving force in the economy.

ening of money market conditions in late June and in July.

This rapid growth of the monetary aggregates in Germany and the large acquisition of foreign exchange by the Bundesbank can be seen as part of an international effort to stabilize exchange rates, especially after the Louvre Accord of February 1987.[17] The Bundesbank's efforts to reduce tension in the foreign exchanges were most evident in its interest rate policy, which contributed to a significant widening of interest differentials in favor of U.S. dollar assets in 1987. The trend toward closer monetary cooperation has also been seen within Europe. The Basle-Nyborg agreement of September 1987 contained provision for joint financing of intramarginal intervention in the European Monetary System (EMS), and during the last two months of 1987, the Bundesbank twice participated in concerted interest rate actions with other European central banks.

The greater weight placed on external considerations in the implementation of monetary policy, however, predated the formal attempts at exchange market stabilization and reflected a growing awareness of the limited policy independence that the adoption of flexible exchange rates affords policymakers. The experience since the breakdown of the Bretton Woods system has illustrated that divergent policies in major countries can produce substantial and protracted deviations of real exchange rates from some notional equilibrium indicated by fundamentals. Such misalignments of exchange rates can be costly in terms of resource misallocation and uncertainty. Moreover, the sensitivity of exchange rates and interest rates to new information (or rumors) can result in unexpected portfolio shifts and consequent problems for short-run macroeconomic management. Problems of this sort are of particular concern in Germany: first, the openness of the economy renders it susceptible to exchange rate misalignments, and, second, the prominent role of foreign investors in the German financial markets and of the deutsche mark in international finance means that international financial turbulence is quickly transmitted to the German financial markets.

Indeed, the influence of developments in the international financial markets on Germany has grown markedly in the 1980s as foreign investors have come to play a major role in the German capital markets. Capital flows in and out of Germany have been relatively free from restrictions for quite some time.[18] With the growing integration of international financial markets and the recent moves of the authorities to strengthen the competitiveness and the international stature of the German financial markets, the role of foreign investors in the German capital market has become much more important in recent years.[19] They now account for a substantial share of the demand for domestically issued bonds and equities (Table A16), in addition to taking up the major part of deutsche mark bonds issued by nonresidents. Issues of deutsche mark bonds by nonresidents have also increased substantially and German investors have been showing greater interest in foreign currency financial assets.

The interaction between international financial market developments and the domestic economy over the past few years can be traced through the recent history of the capital markets, capital flows in the balance of payments, and the foreign exchanges. The volatile conditions in world financial markets were manifested in Germany in a number of ways. First, and most prominent, have been the sharp movements in exchange rates, which have created a great deal of uncertainty for producers, particularly in the export-oriented manufacturing sector. Second, changing interest rates and exchange rate expectations have motivated pronounced changes in international capital flows. Thus, after a period of strong net long-term capital inflows, beginning in the first quarter of 1986, the pattern reversed after mid-1987 as foreign investors shifted out of deutsche mark assets. Third, reflecting their uncertainty about future interest rate developments, domestic nonbank investors were reluctant to place funds in the domestic bond market through 1986 and 1987, despite a steep yield curve that favored longer-term investments.

There is widespread agreement that the flexible implementation of monetary policy by the Bundesbank has helped relieve tension in the exchange markets, and, indeed, during much of 1987, the exchange rate was relatively stable. This course of action has not,

---

[17] The Bundesbank acquires foreign exchange by "active" intervention (i.e., buying foreign currencies) and "passive" intervention (i.e., refraining from its normal practice of selling its regular foreign exchange receipts in the market).

[18] See Deutsche Bundesbank (1985b) for a review of the history of capital controls in Germany.

[19] With effect from August 1984, a 25 percent withholding tax on interest payments to foreign holders of bonds issued by German residents was abolished; this had been the last significant regulatory disincentive to capital flows into Germany. Over the ensuing two years, the Bundesbank took a number of measures to liberalize the rules on the range of financial assets that could be issued in the German financial markets—beginning in May 1985, floating rate notes, swap-related bonds, and zero coupon bonds were permitted and, as of May 1986, the Bundesbank lifted its objections to the issuance of certificates of deposit. There was also a liberalization of rules governing the activities of foreign financial institutions with a presence in the German market. From May 1985, such institutions were permitted to lead-manage issues of foreign deutsche mark bonds and, in June 1986, foreign banks were admitted to the federal bond consortium. In 1987, the period of notice that issuers of foreign deutsche mark bonds are required to give the Bundesbank before a new issue was reduced to one full business day, compared with the previous requirement of two weeks' notice.

however, been without cost; the economy, now much more liquid, is more susceptible to inflationary shocks than it was a few years ago, and some are concerned about damage to the credibility of the Bundesbank. Moreover, there are limitations to what can be achieved by monetary policy on its own. In particular, the Bundesbank does not hold the view that monetary policy can have sustainable real effects or, more specifically, influence real exchange rates over the medium term. Indeed, it has emphasized, that developments during 1986 and 1987 should not be construed as a change in its view of the appropriate and limited role of monetary policy. Rather it sees the low inflation rates in 1986 and 1987 and the lack of an immediate inflationary danger as having provided scope for a more flexible implementation of monetary policy in the short term that was not incompatible with its views about the appropriate role of monetary policy over the medium term.

Two illustrations of the limitations of monetary policy actions are apparent from recent experience. First, the eight months of relative stability in the foreign exchanges following the Louvre Accord saw substantially higher interest rate differentials in favor of the U.S. dollar. The scale of these differentials was indicative of market expectations of significant future exchange rate movements. These expectations were realized, at least in part, as the dollar declined sharply in value following the worldwide fall in equity prices in October 1987. Reflecting the developments in the foreign exchanges in the first two months and again in the last three months of the year, the deutsche mark appreciated by 22¾ percent against the U.S. dollar and by 3½ percent in nominal effective terms during the course of 1987 (Table A17).[20] Second, while the Bundesbank has been able to influence short-term interest rates, it has had much less influence on longer-term rates, which are more important for the interest-sensitive components of expenditure.

The continued turbulence in the foreign exchanges, which has primarily revolved around the U.S. dollar, has underlined in Europe the benefits of the exchange rate mechanism (ERM) of the EMS. With the greater convergence of inflation rates in the participating countries, attention is being increasingly focused on issues related to the further development of the EMS.

The issues raised in this overview are examined below in greater depth. Developments in the monetary aggregates and interest rates are analyzed first. The prominence of international influences in this discussion prompts a more detailed look at international capital flows. This, in turn, leads back to the subject of Bundesbank intervention activities, which have been a significant factor in the development of the monetary aggregates. Finally, there is a brief digression on issues related to the future development of the EMS.

Monetary Developments, 1986–88

In December 1985, the Central Bank Council of the Bundesbank announced a target range of 3½–5½ percent for the growth of central bank money from the fourth quarter of 1985 to the fourth quarter of 1986. This range was set to be consistent with a potential growth rate of 2½ percent and an underlying inflation rate of 2 percent. However, during 1986, for the first time since the inception of the target range in 1979, the growth of central bank money exceeded the upper limit of the target range; indeed in no quarter of the year did the growth rate of central bank money come within the target range established. The principal factor complicating the Bundesbank's control of central bank money during the year was the situation in the foreign exchanges. Only in the second quarter was the Bundesbank a net seller of foreign exchange, as the deutsche mark hovered around its lower intervention point in the EMS following the April realignment. For the year as a whole, however, the Bundesbank increased its foreign exchange holdings by DM 9 billion in an attempt to smooth the upward path of the deutsche mark (Table A18);[21] some of the liquidity effect of this intervention was offset, principally through a reduction in the volume of securities' repurchase agreements.[22]

In December 1986, the Central Bank Council of the Bundesbank established a target range of 3–6 percent for the growth of central bank money during 1987. Thus, the midpoint for the target range for 1987 was the same as it had been in 1986, reflecting an unchanged view on the medium-term growth potential and inflation, but the range had been widened. In widening the target range, the Bundesbank took into account, on the one hand, the fact that the monetary target had been exceeded by a large margin in 1986, and it therefore wanted to allow room for some compensation

---

[20] Most of the appreciation was reversed in the first nine months of 1988. In September 1988, the nominal effective value of the deutsche mark was only ¼ of 1 percent higher than in December 1986. Over the same period, the deutsche mark appreciated by 6¾ percent against the U.S. dollar.

[21] The data on foreign exchange flows in Table A18 are based on monthly average data (to correspond with the central bank money data) and thus can differ from changes in the external position of the Bundesbank in the external accounts; the latter reflect the actual flow during the period in question.

[22] Open market operations in securities' repurchase agreements are the principal tool used by the Bundesbank for guiding money market conditions on an ongoing basis. See Deutsche Bundesbank (1983, 1985c, and 1988e, p.11) for further information on securities' repurchase agreements. Deutsche Bundesbank (1987a) contains a detailed discussion of the various monetary policy instruments used in Germany.

in the growth of central bank money in 1987, should market conditions permit. On the other hand, it was realized that the factors that had complicated monetary management in 1986, particularly the external factors, might again make it necessary to allow growth of central bank money to exceed the 4½ percent growth rate suggested by the underlying growth of nominal productive potential.

In the event, external considerations dominated monetary policy in 1987. Turbulence in the foreign exchanges in the early part of the year, stemming from tensions prior to the EMS realignment in mid-January and from a weak dollar, prompted large-scale intervention. The situation repeated itself in the aftermath of the October stock market disturbance, as pressure on the dollar resumed, once again accompanied by tension in the EMS. Taking the year as a whole, the net flow of foreign exchange into the Bundesbank was about 4½ times that in 1986, a year in which exchange market pressures were also prominent (Table A18). This foreign exchange inflow was more than twice the level of the increase in base money during the year as the Bundesbank sterilized part of the intervention by increasing reserve requirements, reducing rediscount quotas, and lowering the volume of open market securities' repurchase agreements.

Intervention is not the only avenue through which external considerations have influenced monetary conditions. Over the past two years, the Bundesbank has set its interest rate policy with a view to moderating pressures on the deutsche mark; it has then had to support its chosen operating rates by supplying the amount of liquidity required by the market. From this perspective, whether such support came through intervention, repurchase transactions, or other avenues made little difference.[23] Thus, although intervention in 1987 was concentrated in the first and last quarters, central bank money grew significantly above target in every quarter. Indeed, the Bundesbank took the opportunity to sell some of its foreign assets in the third quarter, when pressure on the dollar abated, and an increase in the volume of repurchase transactions was used to keep the market supplied with sufficient liquidity to maintain the low level of interest rates.

While the rate of growth of central bank money during 1987 was similar to that of 1986, the same was not true for the growth of its components. During 1986, currency outstanding grew at the same pace as central bank money, suggesting that the other components of M3 were growing at a similarly fast pace; indeed, the rate of growth of M3 (7¼ percent) was not much lower than that of central bank money. During 1987, however, the growth of currency outstanding jumped to 10¼ percent (7¾ percent in 1986), while the noncurrency components of M3 grew at a rate of 5½ percent (7 percent in 1986). The reasons behind the acceleration in the growth of currency are not fully understood; undoubtedly the reduction in short-term interest rates was a factor, but this does not explain why a stronger demand for liquidity should be reflected in currency and not in sight deposits, which grew more slowly.[24] It is possible, however, that at least part of the increased demand for deutsche mark currency reflected further substitution of the deutsche mark for the U.S. dollar as the preferred parallel currency in some East European economies. Outside of bank notes, however, currency substitution does not appear to have been a significant factor in the rapid growth of monetary aggregates in Germany. Deutsche mark deposits held by nonresidents are not included in the German money supply and holdings of external foreign currency deposits by German residents are relatively small (less than 1 percent of M3).[25]

Because of the large weight of currency in central bank money and also the fact that the rapid growth of currency may not have accurately reflected currency holdings of German residents, it would appear that central bank money growth overstated somewhat the pace of monetary expansion in 1987. Both M3 and monetary assets excluding currency, however, also continued to grow at a fast pace relative to nominal income growth; although both these aggregates grew at a slower rate than in 1986, there was also a considerable slowdown in the growth of nominal income (from 5½ percent to 3¾ percent, on a through-the-year basis).[26] As a result, the velocity of M3 declined markedly in 1987.

---

[23] This is not meant to imply that foreign exchange intervention is irrelevant. It may alter outcomes, particularly over shorter periods, through the signals it sends to market participants. Moreover, when the scale of intervention is large, the ability of the Bundesbank to manage money market conditions is reduced as it is forced to cut back the scale of securities' repurchase agreements, which are its principal and most flexible tool for guiding the money market on an ongoing basis.

[24] It is worth noting, however, that the difference in the growth rates of sight deposits and currency in 1987 was almost entirely attributable to the slow growth of sight deposits in the fourth quarter of the year.

[25] In principle, holdings of deutsche mark bank notes by foreign residents would also be excluded from the German money supply but it is not possible to identify holders of bank notes. Holdings by German residents of foreign currency deposits with the German banking system are included in the German money supply.

[26] While the growth of nominal demand did not slow down in 1987, M1 (which may be a more appropriate variable to focus on from a transactions perspective) also grew at a similar pace to that of the previous year. Nominal domestic demand grew by 4½ percent during 1987 while M1 grew at about 8½ percent. Even if one prefers to focus on sight deposits because of the problems in interpreting currency developments, the same picture of relatively easy monetary conditions is apparent since sight deposits grew at a rate of 7½ percent in the course of 1987.

In interpreting the pace of monetary growth in 1986 and 1987, one should also bear in mind the sharp increase in the external deposits of German nonbanks over this period. The growth of these deposits was particularly fast during 1986 (178 percent from December to December).[27] In part, this growth reflected substitution of external deposits for short-term bank bonds in the portfolios of nonbanks;[28] however, even allowing for this substitution, the monetary aggregate M3 understated the growth of liquidity in the economy over this period by over ½ of 1 percentage point (in comparison to an aggregate including external deposits and short-term bank bonds). In 1987 the demand for deutsche mark credits on the Euromarkets declined somewhat and the external deposits of residents grew during the year (December to December) at the much slower rate of about 17 percent; while this was not a low growth rate, it was not high enough to alter fundamentally the picture presented by the domestic monetary aggregates.[29] The authorities are continuing to monitor these deposits, because they represent a significant source of liquidity not captured by the conventional measure of broad money.[30]

A number of factors were behind the fast growth of the monetary aggregates in 1986 and 1987. The boost given to real disposable income by the changes in the terms of trade and the associated fall in domestic prices supported a sharp rise in the savings rate. At the same time, there was a pronounced shift toward liquidity preference in the economy. Lower short-term interest rates in 1986 and 1987 undoubtedly encouraged greater holdings of currency and demand deposits; however, this seems to have been only part of the story as, despite a sharp steepening of the yield curve (as long-term rates declined relatively little on balance during 1986 and 1987), households were reluctant to invest in the bond market. Enterprises also decided to strengthen their liquid positions, and business-fixed investment did not respond vigorously to the improved profit situation.

Given the uncertainty in financial markets in the last quarter of 1987, the Bundesbank postponed a decision on the monetary target for 1988 from December, when the target is usually set, to January. At its meeting on January 21, 1988, the Central Bank Council decided to retain the same monetary target as for 1987; however, the targeted aggregate was changed from central bank money to M3.

The Bundesbank has indicated that the change in the targeted aggregate does not reflect any change in the philosophy governing monetary policy; over time the two aggregates (central bank money and M3) have tended to move in parallel.[31] Indeed central bank money is essentially a weighted average of the different components of M3; however, since these weights differ considerably by component (with heavier weights for the more liquid components and a particularly heavy weight for currency), special factors affecting one of these components can result in significant divergence in the growth rates of the two aggregates in individual years. During 1987, for example, central bank money grew at over 8 percent while M3 grew at about 6 percent, with the difference attributable principally to the rapid growth of currency. The Bundesbank has indicated that the change to an M3 target was partly influenced by the fact that M3 is more easily understood in international markets. Central bank money, however, has also tended to grow faster than M3 at times of low interest rates, a steep yield curve, and upward pressure on the deutsche mark.[32] All these factors had been present in the recent past and, given the prospect that they might well continue in the near future, it was hoped that M3 would provide a clearer guide to the underlying thrust of monetary policy.[33]

During the first half of 1988, the monetary aggregates continued to grow rapidly, with M3 in June being 7½ percent (on a seasonally adjusted annual basis) above the level of the fourth quarter of 1987[34] as the market

---

[27] This calculation is based on data for deposits placed by nonbank German residents with external branches and affiliates of German banks.

[28] Short-term bank bonds were made subject to reserve requirements with effect from May 1986. This eliminated one of the principal attractions of such bonds for issuers and investors and precipitated a sharp decline in the value of short-term bank bonds outstanding. For further discussion of this and other factors behind the strong growth of external deposits in 1986, see Deutsche Bundesbank (1988a) and pp. 17–19 of this section.

[29] In comparison to an aggregate including these external deposits and short term bank bonds, M3 understated liquidity growth by about ¼ of 1 percent during 1987.

[30] External deutsche mark deposits of German residents were equivalent to about 5 percent of M3 at the end of 1987.

[31] For a discussion of the change in the targeted aggregate, see Deutsche Bundesbank (1988c).

[32] See Kremers (1988), where the differences in the behavior of central bank money and M3 are examined econometrically. The results confirm that, in the long term, the two aggregates grow in tandem but that there can be significant short-term differences in their growth rates.

[33] This, however, may not be sufficient grounds for preferring M3 over central bank money as a monetary target. Kremers (1988), drawing on Argy (1983), analyzes the desirability of alternate monetary targets in the context of an open economy subject to different types of shocks. The choice of target variable should be influenced by the controllability of the alternative aggregates and the stability of their respective links with the economic variables in which the policymakers are ultimately interested (e.g., prices and output). Leaving aside these considerations, it can be said that for financial shocks involving portfolio shifts within the components of M3, M3 is likely to be a better target than central bank money. For other financial shocks and for real shocks, however, the determination of which aggregate is a superior target is somewhat more complex.

[34] Through June central bank money was growing at an annual rate of 8¼ percent.

was provided with sufficient liquidity to sustain the interest rate measures taken at the end of November and in early December.[35] Over this period currency and sight deposits continued to be the principal driving force behind the rapid monetary expansion; however, the Bundesbank took a number of steps in late June and July to tighten monetary conditions, which had the effect of bringing growth back toward the target range.[36]

Interest Rate Developments

Interest Rate Policy

In its execution of monetary policy, the Bundesbank is guided principally by money market conditions, and its monetary policy objectives are most immediately reflected in its interest rate policy. Interest rate policy in Germany during 1986, and particularly in 1987, was very closely linked to external considerations. In the first few months of 1986 the Bundesbank moved its rate on open market repurchase agreements down by about ¼ of 1 percentage point, and in March the discount rate was reduced by ½ of 1 percentage point to 3½ percent. Reflecting these measures, and rate reductions in the second half of 1985, the discount and repurchase rates were over ½ of 1 percentage point lower on average in 1986 than in 1985. In early 1987, the Bundesbank again began to push short-term interest rates lower. Following the EMS realignment in January, the Bundesbank cut its discount rate from 3½ percent to 3 percent on January 23 and other key operational rates were reduced by similar amounts, including the rate on open market repurchase operations (Table A19 and Chart 7).[37] In mid-May, the Bundesbank reduced the rate on securities' repurchase agreements by a further ¼ of 1 percentage point to 3.55 percent.[38]

---

[35] Rediscount quotas were reduced at the beginning of February 1988; this was a technical move to allow the Bundesbank greater influence on money market conditions through repurchase operations.

[36] In September, M3 was 6½ percent above the target base on a seasonally adjusted annualized basis, while central bank money had increased by 7½ percent over the same period.

[37] The Lombard rate was reduced from 5.5 percent to 5 percent. In February and March, the Bundesbank conducted four-week repurchase agreements at a fixed rate of 3.8 percent, compared with the 4.35 percent in effect for most of 1986. (Repurchase agreement rates had been somewhat higher toward the end of 1986 and in the one agreement conducted in January 1987.) The rate at which the Bundesbank offered treasury bills to the banks (normally for a three-day period) was lowered from 4 percent to 3.5 percent; this rate acts as a floor to the money market.

[38] At the same time as the repurchase rate was reduced, the three-day treasury bill rate was lowered to 3.2 percent.

Chart 7. Federal Republic of Germany: Interest Rate Developments

(In percent a year)

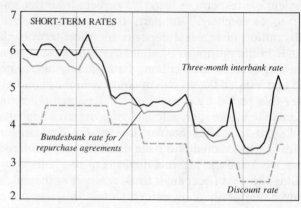

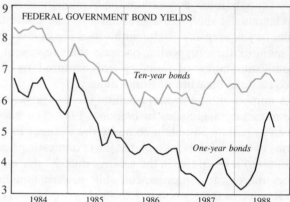

Source: Deutsche Bundesbank, *Monthly Report*, various issues.

During the early summer, when pressure on the dollar declined, money market conditions tightened modestly as the Bundesbank used the opportunity provided by the relative strength of the dollar to sell some of the dollars it had acquired earlier in the year, and toward the end of July it effected a small increase in the securities' repurchase rate (from 3.55 percent to 3.60 percent). In view of the upward trend in international interest rates in the late summer and early fall, and its implication for domestic financial markets, the Bundesbank raised the rate on repurchase agreements further by ¼ of 1 percentage point in three steps between September 23 and October 14.[39] While there was a jump in money market rates in Frankfurt during this period (by the middle of October three-month money was trading at about 1 percentage point

---

[39] On September 23, the rate was raised from 3.6 percent to 3.65 percent, followed by a move to 3.75 percent on October 7 and 3.85 percent on October 14. These were the actual allotment rates under minimum interest rate tenders. (The minimum interest rates quoted by the Bundesbank were 3.5 percent on September 23 and 3.6 percent at the following two tenders.)

above its end-of-September level), it is difficult to attribute much of the jump to changes in the supply of liquidity to the money market as there is no indication that banks were short of funds during the first two weeks of October.[40] While a shortage of liquidity on the money markets does not seem to have been responsible for the rise in the three-month money market rate in the first half of October, there was considerable pressure coming from developments in the international markets as short-term U.S. dollar rates had moved up sharply since August; by the middle of October, the short-term interest differential (three-month London interbank offered rate (LIBOR)) in favor of the dollar had risen a full percentage point from the levels in July and August.

In the midst of the stock market turbulence in mid-October, pressure on interest rates on world markets eased as portfolio preferences shifted toward bond market investments, and stock market developments were taken as presaging a slowdown in the world economy. The downward push imparted to German interest rates was reinforced by the Bundesbank as it, like other central banks, took action to ensure that financial markets had adequate liquidity.[41]

In late October and early November, pressures again mounted on the foreign exchanges, both in the dollar market and within the EMS. In response, on November 5, the Bundesbank took a concerted interest rate action with the Banque de France. In Germany, the Lombard rate was lowered from 5 percent to 4.5 percent, and it was announced that the next repurchase operation (on November 11) would take place at a rate of 3.5 percent; at the same time, the Banque de France raised its intervention rate in the money market by ¾ of 1 percentage point. On November 24, the Bundesbank further lowered its repurchase rate to 3¼ percent, and on December 2, the Bundesbank reduced the discount rate from 3 percent to 2½ percent[42] as part of a joint action with the central banks of Austria, Belgium, France, the Netherlands, Switzerland, and the United Kingdom.

---

[40] First, as the repurchase agreements over this period were conducted as minimum interest tenders, banks that bid at or above the allotment rates set by the Bundesbank were provided all the funds they bid for at the allotment rates. Second, there was not a sharp jump in money market rates for maturities of two months and less; these rates moved up moderately, and only by a little more than the increases in the repurchase rate. As a result, an unusually large gap opened up between these rates and the rate on three-month money.

[41] On October 19, the Bundesbank moved substantial federal government deposits from its own books to the banking system; on October 20, it reduced the repurchase rate to 3.8 percent (compared with 3.85 percent the previous week); and the following week it supplied liquidity to the market through foreign exchange swaps.

[42] This reduced the discount rate to an historically low level. The previous low point was a rate of 2¾ percent, which was in effect for only a few months in 1959.

During the first half of 1988, the Bundesbank continued to supply the liquidity needed to support the low level of official interest rates; however, as the dollar began to appreciate in midyear, it took the opportunity to tighten monetary policy, while at the same time, helping to moderate the downward pressure on the deutsche mark. The repurchase rate was raised from 3¼ percent to 4¼ percent in four steps between June 22 and August 3, and the discount rate was raised from 2½ percent to 3 percent with effect from July 1 and to 3½ percent with effect from August 26.

Developments in Market Rates

The Bundesbank's interest rate policy has supported a sustained downward movement in money market interest rates from early 1985 through early 1988, with the exception of a few brief episodes (Chart 7). The average rate for three-month money fell from 5.4 percent in 1985 to 4.6 percent in 1986 and 4 percent in 1987 and reached its low point of 3.3 percent in February 1988, which was very low by historical standards (Chart 8). Reflecting principally the interest policy measures taken by the Bundesbank in the summer of 1988, the three-month money market rate rose back to 5 percent in September 1988.

The interest rate policy of the Bundesbank helped reduce tensions in the foreign exchanges although its contribution in this respect cannot be simply read from

Chart 8. Federal Republic of Germany: Interest Rates in Historical Perspective

(In percent a year)

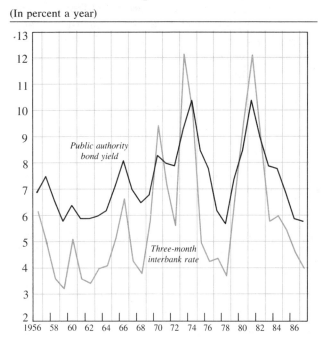

Source: Deutsche Bundesbank, *Monthly Report*, various issues.

III • DEMAND MANAGEMENT POLICIES AND FINANCIAL MARKET DEVELOPMENTS

what has happened to short-term interest differentials between U.S. dollar and deutsche mark denominated assets. Indeed, during the course of 1986, short-term interest differentials in favor of U.S. dollar denominated assets narrowed markedly (Chart 9), as the large depreciation of the dollar that had commenced in spring 1985 had apparently reduced expectations of future depreciation. In early 1987, however, there seems to have been a fundamental reappraisal of the market's view of the dollar's prospects and the dollar came under substantial pressure in the foreign exchanges. The period of exchange rate calm in the eight months following the Louvre Accord (February 22, 1987) was accomplished at least in part through a significant widening of interest differentials, supported by a decline in short-term deutsche mark interest rates and a rise in the corresponding dollar rates.[43] After a period of dollar strength in the summer, differentials in favor of the dollar widened further and by the end of the year they were greater than at any time since 1984. Although there was some decline from their end-of-1987 level, short-term interest differentials in favor of the dollar remained high through September 1988, particularly in comparison with the situation at the end of 1986.

The Bundesbank has clearly been able to exert considerable influence on shorter-term interest rates, though even here its influence has been constrained by developments in the world financial markets, as was apparent in October 1987. A marked feature of developments, however, has been the limited ability of the Bundesbank to influence long-term rates; the yield on long-term (ten-year) government bonds did not on balance show a downward trend from the end of 1985 to the spring of 1988, while short-term rates declined significantly.[44] Over this period, there was a pronounced weakness of domestic demand for bonds.[45] This was compensated in 1986 and early 1987 by strong

---

[43] There was also substantial intervention in support of the dollar by many central banks. Worldwide accumulation of official U.S. dollar reserves was not much smaller than the U.S. current account deficit in 1987.

[44] Although the average government bond yield, based on all bonds with a remaining maturity of more than three years, did show some downward movement, it reflected the influence of Bundesbank action on shorter-term bond yields within this category.

[45] Bond purchases of domestic nonbank investors, which amounted to DM 12 billion in 1986 and DM 34 billion in 1987 (Table A20), remained significantly below the average level of 1980–85 (DM 43 billion). Thus, despite the increase of nominal household income of about 10 percent over 1986–87, an increase in the savings rate, and a sharp steepening of the yield curve, nonbank residents reduced the nominal value of bond purchases compared with earlier years. Moreover, a large proportion of the bond purchases by nonbank residents in 1986–87 went into bonds denominated in foreign currency and, when one takes this into account, the decline in the purchase of deutsche mark bonds was even more pronounced. See Deutsche Bundesbank (1988d) for further discussion of the investment behavior of domestic nonbanks in the bond market.

Chart 9. Federal Republic of Germany: Short-Term U.S. Dollar and Deutsche Mark Interest Rates

(In percent a year)

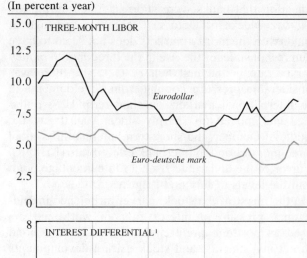
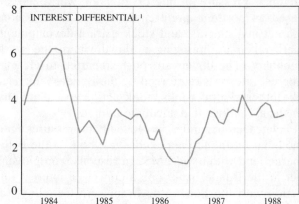

Source: International Monetary Fund, *International Financial Statistics*, various issues.
[1] U.S. dollar rate minus deutsche mark rate.

long-term capital inflows that facilitated some decline in long-term yields in Germany between the end of 1985 and May 1987. But, when the direction of long-term capital flows was reversed in mid-1987, encouraged by higher U.S. dollar interest rates and greater confidence in the dollar, there was significant upward pressure on long rates in Germany. The behavior of domestic investors seems to have reflected fears of rising interest rates in the future. In part, this may have resulted from concern about the inflationary potential of rapid monetary growth and the impact on financial markets of the prospective increases in fiscal deficits. It may also have reflected a psychological reaction to the fact that bond yields were low by historical standards.[46]

That the markets were expecting future rises in interest rates was suggested by the steepness of the yield curve, itself a reflection that the pronounced fall

---

[46] Chart 8 compares the average public sector bond yield in 1987 with yield levels over the period 1956–86. Only in one year (1978) was the bond yield lower than in 1987 and then only by a small amount.

in short-term rates was not accompanied by marked declines in long-term rates. It was also seen in the pattern of borrowing. The stock of outstanding short-term credit extended by banks to the private sector fell at an annual rate of 3 percent from the fourth quarter of 1985 to the fourth quarter of 1987. Over the same period, the stock of long-term credit extended by the banks continued to grow at a pace of over 6 percent. This occurred in spite of the much larger decline in short-term rates, suggesting that borrowers perceived long rates to be low and were taking advantage of the opportunity to transform short-term debt into longer-term debt.

Capital Market Developments and the Capital Account of the Balance of Payments

The linkages between the German capital markets and the international financial markets have clearly been an important influence on monetary policy and financial market conditions in recent years. This has perhaps been most apparent in the shifting pattern of capital flows in the balance of payments. In 1986, in contrast to the long-term capital outflows that have been more typical in the 1970s and 1980s, there was a substantial net long-term capital inflow of DM 33 billion (Table A21). This inflow plus the current account surplus of DM 85 billion was offset by an unusually large net short-term outflow (including unclassifiable items) of DM 112 billion (Table A22) and a small increase in the net foreign assets of the Bundesbank (DM 6 billion at transactions values). In 1987, reflecting a significant decline in foreign investment in Germany, a net long-term capital outflow of DM 24 billion helped to offset part of the current account surplus of DM 81 billion. The net short-term capital outflow, which was much reduced from 1986, contributed a further DM 16 billion. Over half of the current account surplus in 1987, however, was reflected in a buildup in Bundesbank foreign assets amounting to DM 41 billion (at transactions prices). In the first nine months of 1988, long-term capital outflows exceeded the current account surplus and with short-term capital also leaving Germany, the Bundesbank reduced its net foreign assets by DM 35 billion.

Long-Term Capital

The sharp changes in the pattern of long-term capital flows over the past few years were concentrated principally in portfolio investment, that is, investments in the German bond and equity markets by foreign investors and the corresponding investments in foreign securities by German residents.

The Bond Market. In 1986, nonresident net purchases of domestically issued bonds jumped to DM 59 billion from DM 31 billion in 1985 (Table A20). In 1987, they fell to DM 35 billion; indeed, in the second half of the year, foreigners did not purchase (net) any bonds on the German financial markets. The fall in foreign purchases in 1987 was even more dramatic if one takes into account foreign holdings of official borrowers' notes.[47] In the first nine months of 1988, nonresident investors continued to stay away from the German bond market.

The interest of foreign investors in the German bond market in 1986 had been sparked by exchange rate expectations and incipient upward pressures on yields in the German market as the interest of German residents in the bond market cooled.[48] Also, the slope of the yield curve in Germany, in conjunction with expectations of an appreciating deutsche mark, seems to have encouraged foreign investors to take up short-term deutsche mark liabilities (or reduce deutsche mark deposits) on the Euromarkets and use the proceeds to invest in longer-term bonds on the German market. There were substantial foreign purchases of German bonds in the first half of 1987, particularly in January when tension in the EMS and pressure on the dollar stimulated expectations of exchange rate gains and the prospect of falling interest rates promised capital gains; however, in the second half of the year foreign investors retreated from the German bond market. Apparently, foreign investors feared possible capital losses resulting from rising bond yields, and changed expectations with respect to exchange market developments made the prospective gains from such yield curve arbitrage less attractive than previously when converted back to the investor's home currency. The behavior of German investors in foreign bond markets mirrored that of foreign investors in the German bond market and reflected similar influences.[49]

---

[47] These borrowers' notes were quite popular with foreign investors prior to the abolition of the coupon tax in 1984 as they were not subject to the tax. Since the abolition of the tax, foreign investors have been reducing their holdings of such notes as they have come due and replacing them with bonds in their portfolios. Consolidating official borrowers' notes with bonds, nonresident net purchases of bonds declined from DM 54 billion in 1986 to DM 23 billion in 1987.

[48] In the absence of the increased foreign demand, the reduced interest of domestic residents in the bond market would have put upward pressure on yields. To this extent, the increased foreign demand reflected the process by which interest differentials were kept consistent with exchange rate expectations and factors affecting the degree of substitutability between deutsche mark and U.S. dollar assets.

[49] Thus, domestic demand for foreign currency bonds was strong over the period March to July 1987, as foreign exchange markets stabilized and investors were attracted by the high yields on these bonds. With the unsettled conditions in foreign exchange markets in the last quarter of the year, resident net purchases of foreign currency bonds dropped sharply. As relative calm returned to the foreign exchanges during most of the first half of 1988, foreign

III • DEMAND MANAGEMENT POLICIES AND FINANCIAL MARKET DEVELOPMENTS

The sharp change in the attitude of foreign investors had a significant impact on bond market yields during 1987. In the first few months, yields declined as foreign interest continued to be strong but by May they had bottomed out. Over this period, the decline in yields was a good deal stronger for short-term bonds; while the yield on one-year bonds in May was a full percentage point below its end-of-1986 level, the yield on ten-year bonds fell by about 40 basis points over the same period. With the retreat of foreign investors during the remainder of the year and with domestic investors unwilling to take up the slack, yields increased sharply; since monetary policy was holding rates down at the short end of the market, this was reflected in a further steepening of the yield curve. In the second half of the year, the yield curve was steeper than it had been at any time over the previous twenty years, with the exception of the first half of 1975. The steep yield curve persisted during the first five months of 1988, but as short-term rates moved up in June and July, there was, in comparison, relatively little impact on long-term rates, and by the end of September the yield curve was exhibiting a more normal slope.

The Share Market. Foreign investors have been an important source of demand for German equities in recent years (Table A16). In 1986, net foreign purchases of German equities of DM 15 billion accounted for close to half of all purchases on the German stock markets and were equivalent to 93 percent of the new issue of domestic shares during the year; thus, on a net basis, acquisition of equities by domestic residents was almost entirely focused on foreign stocks. However, in 1987, nonresidents reduced their holdings of German stocks (reflecting large net sales in the fourth quarter), and the gap in the market for domestic equities was covered by domestic residents, whose net purchases of foreign equities fell sharply. Nonresidents continued to reduce their holdings of German equity in the first six months of 1988. These were taken up by German nonbanks, who also resumed significant purchases of foreign equities.

Foreign investors are said to play an important role in influencing price developments in the German equity market. While this may seem somewhat surprising in light of the relatively small share of the total amount of German equity held by nonresident investors (Table A23), foreign investors are particularly important in terms of turnover.[50] Thus, in 1986, a period of strong foreign demand, share prices were on average 41 percent above their average level in 1985 and about 13 percent up during the year;[51] during 1987, however, share prices fell by about one third, as demand from foreign investors weakened. The German market did not have the surge in prices that characterized the New York, London, and Tokyo markets in the first eight months of 1987 and was hit more severely than the other major markets by the sharp decline in share prices worldwide that set in in September but that was particularly steep in October and November. Between August 1987 and February 1988, industrial share prices fell by 22 percent and 23 percent in the United Kingdom and the United States, respectively, and by only 6 percent in Japan; in Germany the decline was 34 percent.[52]

Developments during 1987 cast an interesting light on the factors that have influenced equity prices in Germany. It is rather notable that the decline in stock prices by 14 percent in the first five months of 1987 occurred at a time of strong demand from foreign investors in the German bond market. Among the factors that encouraged foreign investment in the bond market, the hope of declining interest rates and of currency gains would also have been expected to boost demand for German equities. It would appear, however, that a more pessimistic view of the economic outlook and, in particular, concerns about the impact of exchange rate developments on German industry overshadowed these factors. Thus, after May, when foreign investors withdrew from the German bond market as the prospect of currency stability became more accepted, confidence returned somewhat to the stock market and prices increased by 14 percent between May and August. The subsequent decline in share prices, following world equity market developments, was reinforced by the return of exchange market turbulence in the final quarter of 1987 and the accompanying concerns about the future profitability of enterprises.[53] Equity prices on the German market recovered by 23 percent between January 1988 and September 1988, as exchange rate developments boosted

---

currency bonds again became a major focus of bond purchases by German residents.

[50] A very large part of German equity is held by nonfinancial corporations and banks, who tend to view these as longer-term investments. The household sector, which is a relatively small holder of equity by international standards, is also believed to take a longer-term view in its purchases and sales in the equity market.

[51] Share prices in December 1986 were, however, about 10 percent below their peak in mid-April 1986.

[52] In part the sharper fall in German prices is likely to have reflected the impact of exchange market uncertainty on the prospects of export-oriented producers in conjunction with the greater dependence of German producers on the export market (see discussion below). Changes in the market's perception of future profitability of a company, however, are also likely to have a greater impact on share values of more highly leveraged companies, and German enterprises are typically more highly leveraged than those in the United States or the United Kingdom. Of course, this doesn't explain developments in Japan where enterprises are also relatively highly leveraged.

[53] See Deutsche Bundesbank (1988b), pp. 17–18 for a discussion of how export market considerations seem to have affected German equity prices in late 1987 and early 1988.

export market prospects but prices were still 25 percent below their end-of-1986 levels.

Short-Term Capital

The large outflow of short-term capital in 1986, which was the counterpart of the surpluses on the current account and the long-term capital account, was reflected in a substantial buildup of deutsche mark Euromarket deposits and a reduction of Euromarket liabilities of German residents (both banks and nonbanks). There were a number of reasons for the sharp increase in the demand for deutsche mark on the Euromarkets. First, large-scale intramarginal intervention in the EMS resulted in significant withdrawals of Euro-deutsche mark deposits by European central banks.[54] Second, the widening of the German current account surplus and the fact that most of Germany's foreign trade is denominated in deutsche mark elicited a strong demand for deutsche mark trade financing. Finally, as already discussed, nonresidents increased their demand for short-term deutsche mark credit and reduced their deutsche mark deposits to finance purchases of long-term bonds. This strong demand for deutsche mark on the Euromarkets pushed Euro-deutsche mark rates somewhat higher than normal in relation to domestic rates.[55]

The sharp decline in the net short-term capital outflow of both banks and nonbanks in 1987 reflected reduced demand for deutsche mark credits on the Euromarkets as the factors responsible for the strong demand in 1986 were no longer as prominent; this situation persisted in the first half of 1988.[56]

Net Foreign Assets of the Deutsche Bundesbank

A prominent feature of balance of payments developments in 1987 was the role played by changes in the foreign asset position of the Bundesbank, which amounted to DM 41 billion in 1987 compared with DM 6 billion in 1986. These changes in the Bundesbank's foreign asset position reflected two factors—intervention in the foreign exchange market and an accumulation of reserves through autonomous inflows.[57] Accumulation of these autonomous inflows can be thought of as passive intervention, as it is the Bundesbank's normal practice to sell foreign exchange acquired in this way on the market. About half of the increase in the Bundesbank's foreign assets reflected intervention in the EMS. This intervention occurred in two episodes. First, there was obligatory intervention by the Bundesbank prior to the EMS realignment of January 12, 1987. Second, under the joint financing arrangement established as part of the Basle-Nyborg agreement of September 1987, short-term credit was provided through the European Monetary Cooperation Fund (EMCF) to help finance intramarginal intervention in the EMS by other central banks when tensions rose in late October and early November. The other half of the increase in net foreign assets reflected dollar accumulation, resulting particularly from purchases of dollars during the first and last quarters of the year, and from autonomous inflows. During the first half of 1988, the net foreign assets of the Bundesbank fell by DM 13 billion, much of this was in May and June when, with the strength of the dollar, the Bundesbank decided to pass into the market some of the foreign assets it had acquired in 1987. From July to September, with the continuing strength of the dollar, the net foreign assets of the Bundesbank declined by an additional DM 22 billion.

The Future Development of the EMS

The increased convergence of inflation rates within the EMS and the favorable perceptions of the exchange rate stability that the EMS has provided in an environment characterized by significant global exchange market pressures has focused attention on the future development of the system, including the scope for closer monetary integration.[58] In this respect, implications for monetary policy in Germany are of particular interest, given the anchor role that German monetary policy has played in the EMS since its inception.

In 1987, at the Basle-Nyborg meetings in September, a strategy was agreed upon to enhance coordination of policies within the ERM at times of exchange market pressures. This agreement extended the very short-term financing facility of the EMCF to cover intramarginal interventions, subject to two conditions. First, since intramarginal intervention was to be (as it

---

[54] A discussion of intramarginal intervention in the EMS in 1986 can be found on pp. 70–71 of Deutsche Bundesbank (1987b). The corresponding discussion of intramarginal intervention in 1987 is on pp. 64–70 of Deutsche Bundesbank (1988f).

[55] Normally, the three-month Euro–deutsche mark deposit rate on the interbank market lies below the corresponding Frankfurt rate, but for much of 1986 it was higher than the Frankfurt rate. (See Deutsche Bundesbank (1988a).)

[56] First, net intramarginal intervention against the deutsche mark was on a smaller scale in 1987 (see Deutsche Bundesbank (1988c, p. 68)) and given new arrangements in the EMS for financing such intervention (ibid., pp. 64–70), there was no need for other EMS central banks to run down their deutsche mark deposits. Second, the German current account surplus fell in 1987. Finally, as already discussed arbitrage between low Euromarket rates and German bond yields was less prominent in 1987.

[57] Most important among these inflows are interest income and foreign currency exchanged by foreign troops stationed in Germany.

[58] See Ungerer and others (1986), Guitián (1988), and Russo and Tullio (1988) for discussions of recent developments in the EMS.

had been in the past) subject to the approval of the central bank whose currency was being intervened against, access to the very short-term financing facility in these circumstances was not to be automatic (in contrast to the automatic access in the case of obligatory interventions).[59] Second, such access to the financing facility would also be subject to upper limits (again in contrast to obligatory interventions where access is not limited). Despite these limitations, the use of the short-term financing facility for intramarginal intervention represents a significant change, in that intramarginal intervention financed through the financing facility has liquidity effects in the country against whose currency the intervention is taking place.[60] Previously, intramarginal intervention against the deutsche mark had for the most part been financed by a drawing down of deutsche mark deposits held by the intervening central bank in the Euromarkets, and not from reserves they held at the Bundesbank; thus there was no significant liquidity effect in Germany. In addition to intramarginal intervention, the strategy developed at Basle-Nyborg involved greater use both of the fluctuation band for exchange rates to discourage speculators and of coordinated interest rate actions.[61]

In discussions of the future development of the EMS, three issues have been prominent. First is the issue of the so-called asymmetry of the system, that is, the greater pressure to adjust on countries that are following relatively more expansionary policies. This is largely a political issue of the reconciliation of different tolerances for inflation. It can be noted, however, that the substantial convergence of inflation rates that has taken place in the 1980s has been in the direction of lower inflation rates.

Second is the question of how the goal of closer monetary integration can be accomplished and in particular the form a European central bank might take and how such an institution could evolve.[62] Here different views will need to be reconciled about the degree of independence from political institutions that a European central bank should have.

Third, despite the greater degree of monetary convergence that has already taken place among EMS countries, concern still exists that substantially more progress needs to be made in other policy areas before closer monetary integration can be successful. It has been pointed out that fiscal convergence has been less evident and that some member countries have felt it necessary to maintain controls on the flow of capital. Clearly, none of these problems is insurmountable, and indeed progress has been made on the liberalization of capital flows within the EC.[63]

There is, however, a more fundamental economic concern as to whether it is possible for the EMS (particularly if it is moving toward greater fixity of exchange rates and ultimately a common currency) to accommodate substantial differences in potential growth rates among its members. The issue touches particularly on Germany since it has a relatively low potential growth rate (owing to demographic factors) and there have been calls for Germany to increase its growth rate to facilitate the smooth operation of the EMS.

An analysis of this problem (McDonald (1988b)) shows that, if goods are differentiated by country of origin and labor is not fully mobile, there will need to be terms of trade changes between high-growth and low-growth countries. Such changes are real rather than nominal, but they may be easier to effect by nominal exchange rate changes in the short run when certain other nominal variables are insufficiently flexible. As a medium- to long-run proposition, it is not reasonable, however, to expect in a sophisticated economy that frequent use of a nominal variable, such as the exchange rate, to achieve real effects can be successful in the absence of forces in the economy that foster the required real changes. By reducing the rigidities that hamper the operation of such forces, it may be possible to achieve the required terms of trade changes without recourse to nominal exchange rate changes.[64]

Fiscal Policy

Principles

The present fiscal policy in Germany has to be seen in light of the historical experience of the late 1960s

---

[59] Once approval is given for the intramarginal intervention, financing may occur without further approval.

[60] The actual liquidity impact over time depends on the extent to which the intervening central bank repays the credit from the very short-term financing facility in the intervention currency or in other currencies (including the ECU). The central bank providing credit can request repayment in its own currency. Moreover, there is nothing, in principle, to prevent the central bank that provides credit from taking offsetting (sterilizing) actions in its domestic money market.

[61] The period of tension at the end of October and in early November 1987 marked the first use of the joint facility for the financing of intramarginal intervention, in conjunction with the other elements of the strategy worked out at the Basle-Nyborg meetings (i.e., greater use of the band and coordinated interest rate policy).

[62] See Russo (1988) for a discussion of these issues.

[63] See International Monetary Fund (1988).

[64] To the extent that inflation is very low in the slower-growing countries, this may become more difficult as it might require declines in nominal price levels in the faster-growing countries. This aside, there seems little that aggregate demand policies in the low-growth countries can do on a lasting basis to accommodate differential underlying growth rates. To the extent that rigidities impede growth in the low-growth countries, action to reduce these rigidities would reduce the need for terms of trade changes.

Chart 10. Federal Republic of Germany: Selected Indicators of Fiscal Policy

(In percent of GNP)

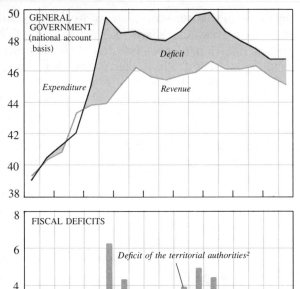
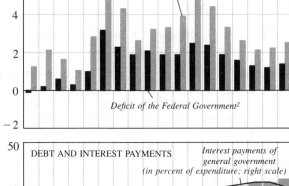
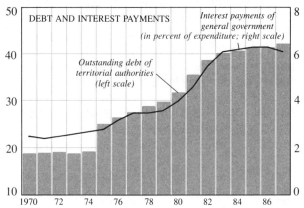

Sources: Statistisches Bundesamt, *Volkswirtschaftliche Gesamtrechnungen* and *Wirtschaft und Statistik*, various issues.
[1] Administrative basis.

and 1970s. During these years, fiscal policy was clearly directed toward supporting economic activity. "Keynesianism" had its debut in Germany during the 1966–67 recession, the first serious slowdown of economic activity in the postwar period. The Government countered this recession with a program of deficit spending during the slump and consolidation of government finances during the upswing that followed.

During the next recession, which was triggered by the first round of oil price increases in 1973, the Government again took measures to boost economic activity; in this cycle, however, it failed to eliminate the ensuing public sector deficits when the economy picked up in 1976 (Chart 10). The main reason for the decision not to consolidate government finances during the upswing was the continuing high level of unemployment. Government support of economic activity throughout the business cycle was considered necessary to boost employment. But the policy was not successful. The effects of further expansionary fiscal measures in 1978, taken in response to domestic and international pressures, reinforced inflationary pressure in the economy. The resistance of monetary policy to fully accommodating the increase in inflation, which resulted from these developments and the second oil price shock in 1979, contributed to the subsequent economic downturn. Fiscal deficits, which had already been increasing steadily since 1977, rose sharply in 1980–81, this time as a result of the slowdown in economic activity rather than expansionary policies. In 1982, the Federal Government changed its fiscal policy stance and introduced an austerity budget that reduced the deficit despite the deepening recession (Chart 11).

The new administration, which came to office at the end of 1982, proclaimed a programmatic U turn (*Wende*)

Chart 11. Federal Republic of Germany: Contribution of Public Consumption and Investment to Growth of Real GNP

(In percent)

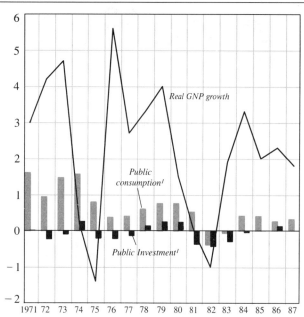

Sources: Statistisches Bundesamt, *Volkswirtschaftliche Gesamtrechnungen* and *Wirtschaft und Statistik*, various issues.
[1] Contributions to growth in percentage points.

in the conduct of fiscal policy. As a first step, it promised "quantitative" fiscal consolidation, that is, the reduction of government expenditures and deficits relative to GNP with a view to reducing or at least stabilizing the ratio of public debt to GNP. This, it was thought, would restore some room for discretionary government spending and release resources from the public sector to the private sector, where they would be used more efficiently for investment and growth. The next step in the new Government's program was "qualitative" fiscal consolidation, which comprised the restructuring of government expenditures and revenues with a view to increasing economic efficiency. Within government expenditures, it was envisaged that the share of subsidies and public consumption would decline while that of investment expenditures would rise. A relative decline in current spending and an increase in capital spending was deemed appropriate in light of the increasing need for environmental protection and basic scientific research. On the government revenue side, a major overhaul of the tax system was planned with a view to reducing distortions and improving work incentives. Tax reform was expected to lower government revenues; it was consequently seen as a second phase of the program, scheduled to take effect only after sufficient progress in quantitative consolidation had been achieved.

The new fiscal policy eschewed the stop-and-go policies of the past in favor of a medium-term framework. This, it was argued, would stabilize public expectations and thus create an environment conducive to an expansion of private activity. In order to ensure that all levels of government—that is, state and local governments, as well as the Federal Government—supported the new fiscal policy, the Financial Planning Council, which coordinates fiscal policy between the federal and state governments and the municipalities, set the target that the rate of growth of nominal government expenditures should not exceed 3 percent a year, a rate considerably below the envisaged rate of growth of nominal GNP.

Achievements

The Period of Quantitative Consolidation

Between 1982 and 1985, the ratio of general government expenditures to GNP declined by 2½ percentage points from almost 50 percent to 47½ percent (Table A24).[65] At the same time, the share of revenues in GNP declined only slightly so that the deficit narrowed from 3¼ percent to 1 percent of GNP. The impressive achievement in the quantitative consolidation, however, was not matched by progress in qualitative consolidation.

On the expenditure side, contrary to the Government's objectives, the share of subsidies in GNP increased (by 11 percent) while the share of public investment in GNP declined sharply (by 18 percent). Success was achieved, however, in reversing the past rapid growth of social transfers and public consumption (expenditures for these purposes dropped in relation to GNP by 9 percent and 3 percent, respectively, between 1982 and 1985). On the revenue side, indirect taxes fell relative to GNP, reflecting the significant share of specific taxes in the total, while direct taxes (excluding social security contributions) rose, largely as a result of fiscal drag; the latter development was also not consistent with the Government's priorities. As a result of the decline in social transfers, however, social security contributions could be reduced relative to GNP (by 3 percent).[66]

The Shift to Qualitative Consolidation

The policy of fiscal consolidation and containment of social expenditures brought the general government deficit in 1985 to its lowest level relative to GNP since 1975. Similarly, the deficit of the territorial authorities and the Federal Government reached a trough in absolute terms. Public deficits, however, were still not low enough to stabilize the ratio of gross public debt to GNP over the medium term.[67] Despite this, the authorities went ahead with a DM 11 billion (0.6 percent of GNP) tax cut in 1986 that was considered to be the first step in a larger reform of the tax system. The next step, in the form of a further tax cut of DM 8.5 billion (0.4 percent of GNP), was planned for 1988, with further tax reform envisaged in the 1990s.

---

[65] The general government consists of the social security system and the territorial authorities. The term territorial authorities refers to the Federal and Länder Governments, the municipalities, the Burden Equalization Fund, the European Recovery Program Fund, and the European Communities' Accounts. While fiscal policy is conducted at the level of the territorial authorities, the following discussion refers to the general government in order to take into account changes in the social security system that were also implemented during the period under review.

[66] Other government revenues, at 4 percent of GNP, were higher in 1982–85 than in the periods before and after (Table A24). This reflected exceptionally high transfers of profits from the Bundesbank to the Federal Government, which contributed to fiscal consolidation (see also below).

[67] With the ratio of the federal debt to GNP amounting to 21.3 percent of GNP by the end of 1985 and an underlying rate of nominal GNP growth of around 4½ percent, the federal deficit consistent with a stable ratio of the federal debt to GNP was slightly below 1 percent of GNP, compared with the actual federal deficit in 1985 of 1.3 percent of GNP. Similarly, with debt of the territorial authorities amounting to 41.2 percent of GNP by the end of 1985, the deficit consistent with the stabilization of this ratio was 1.9 percent; the actual deficit, however, was 2.1 percent.

A gradual approach to tax reduction and reform was chosen so as not to jeopardize further progress in the reduction of public sector deficits that, by the government's own standards, was still necessary.[68]

Beginning in 1985, however, domestic and international pressures for a more expansionary fiscal policy began to build up. On the domestic side, after several years of fiscal austerity, expenditure restraint by state and local governments began to weaken; from an international point of view, the large increase in Germany's external surplus called into question the appropriateness of a fiscal policy geared toward the further reduction of fiscal deficits. In the event, these considerations contributed to a change in the stance of fiscal policy. But in 1986, despite the tax cut and strong expenditure growth at the state and municipal levels, the deficit of the territorial authorities increased only slightly, owing to a significant cutback in the growth of federal government expenditures. Similarly, the general government deficit increased only slightly as social security continued to run large surpluses.

The budgetary plans of the territorial authorities for 1987 foresaw a further modest decline in the aggregate deficit to slightly under 2 percent of GNP.[69] Expenditures of the Federal Government were budgeted to increase by 2.7 percent, up from 1.7 percent the year before, but revenues were expected to increase by 3¼ percent, after 1¾ percent in 1986, allowing the federal deficit to decline by DM ½ billion to DM 22¾ billion (1 percent of GNP).

In the event, the deficit of the territorial authorities increased by a substantially larger amount than expected (Table A25). Altogether, it rose to DM 51 billion (2.5 percent of GNP); of this, DM 28 billion (1.4 percent of GNP) was accounted for by the Federal Government, DM 20 billion (1 percent of GNP) by the state governments, and DM 2¼ billion (0.1 percent of GNP) by the municipalities (Tables A25–A28). Expenditures of the territorial authorities continued to grow at a rate above target for the third year in a row. This was due largely to a strong rise in outlays by the Länder and municipalities for wages, salaries, and social transfers, but also to a sharp increase in the purchases of goods by state governments.[70] In addition to the higher-than-targeted expenditures of the territorial authorities, there was also a substantial revenue shortfall. In particular, as already anticipated in May 1987, tax revenues increased by about DM 10 billion less than projected in the tax estimates of November 1986. Several factors were responsible for the shortfall: most important, nominal GNP was lower than assumed in the tax estimates, and revenues from the corporation tax dropped by 15½ percent from their 1986 level, owing to the reimbursement of taxes received in 1986 in connection with the sale of a major industrial corporation and to lower prepayments of corporation taxes.[71,72] Moreover, other revenues of the Federal Government also fell short of their budgeted level owing to the postponement of the sale of government holdings in Volkswagenwerk AG, after the drop in share prices in October–November, and to lower fees received for credit guarantees by the Federal Government.

The surplus of the social security system declined by DM 2½ billion to DM 7 billion (0.3 percent of GNP), owing largely to a drop (from DM 7½ billion to DM 4¾ billion) in the surplus of the pension fund. This reflected, in part, a reduction in contributions to the Fund by employers and employees, from 19.2 percent to 18.7 percent of gross wage income effective from January 1, 1987; as a result, revenue increased by only about 1½ percent in that year. The number of retirees also rose by about 1½ percent in 1987 and pensions were increased by 3.8 percent effective July 1, 1987. At the same time, however, contributions of pensioners to their medical insurance were raised by 0.7 percentage point[73] so that total expenditures of the pension fund rose by only 3 percent.

Contributions to the unemployment insurance scheme by employers and employees were increased from 4.0 percent to 4.3 percent of gross wage income effective January 1, 1987. But, at 12¼ percent, the rise of expenditures was stronger than that of revenues (10½ percent) so that a deficit of DM ¾ billion emerged. The main reason for the sharply higher expenditure was an increase in unemployment benefits, caused by a larger number of unemployed and an extension of unemployment payments for certain higher-age groups of the unemployed effective July 1, 1987. More short-time workers and a bigger bill for the education and training programs of the unemployment insurance

---

[68] See Federal Republic of Germany (1985, p. 40).

[69] See Bundesministerium der Finanzen, *Finanzbericht 1987* (Bonn, 1986), p. 76.

[70] The 4 percent increase in the labor costs of the states reflected a 3.4 percent increase in public sector wages together with some wage drift. Labor costs of municipalities increased by 5¼ percent owing to the above factors and a 1½ percent increase in the number of employees. Higher current transfers at the municipal level were due to a rise in social aid, inter alia, to the long-term unemployed; in the case of the states, the sharp increase in transfers was chiefly elicited by an increase in support to regions that depend upon the steel and coal mining industries.

[71] Companies are allowed to reduce their prepayments of corporation tax when they expect a decline in profits.

[72] The strong increase in revenues from the wealth tax (23 percent), income tax (7 percent), and the value-added tax (7 percent) allowed total tax revenues to rise by 3½ percent despite the decline in revenues from the corporation tax (Table A29).

[73] Consequently, the amount paid by the pension fund for the health insurance of retirees fell by 8½ percent while the contributions by individuals increased by 4 percent.

system also contributed to stronger expenditure growth.[74] To cover the deficit, reserves of the unemployment insurance fund were used; by the end of 1987, these reserves had declined to about DM 4 billion.

Expenditures of the medical insurance fund had outpaced revenues in 1986, requiring an increase in the rate of contribution from 12.2 percent in that year to 12.6 percent in 1987. This resulted in a small budgetary surplus (of DM ½ billion) in 1987. Nevertheless, reflecting in part the decline in the aggregate surplus of the social security system, the deficit of the general government increased from DM 23½ billion (1.2 percent of GNP) in 1986 to DM 34 billion (1.7 percent of GNP) in 1987 (Table A30).[75]

While 1986 might be described as the year in which quantitative consolidation came to a standstill, 1987 could be characterized as the year in which it was partly reversed. Not only did fiscal deficits increase again in 1987 but decisions were taken that changed the medium-term fiscal outlook. In early 1987, after being re-elected to office, the Federal Government decided upon the long-planned major tax reform package.[76] It was to include a gross tax reduction of DM 42 billion, partly offset by a cut in tax exemptions and a closing of tax loopholes that was expected to raise revenues by some DM 18 billion; thus "net" tax relief was to amount to about DM 24 billion. In February 1987, it was decided that DM 5 billion of this package would be used to increase the tax cuts scheduled for 1988, with the rest to go into effect in 1990. As a result of these decisions, and contrary to earlier intentions, fiscal deficits of the territorial authorities (and the general government) were expected to increase significantly over the next few years.

The Outlook

Budgetary Decisions for 1988

With expenditures budgeted to rise by 2½ percent and revenues to fall by ¼ of 1 percent (owing to the scheduled tax cuts), the federal budget that was passed by Parliament in November 1987 foresaw a deficit of DM 29¼ billion (1.4 percent of GNP). This deficit, though much higher than that originally budgeted for 1987, was broadly unchanged from the 1987 outturn. The deficit was expected to rise to about DM 32 billion in 1990, when the next tax reductions would take effect, but to decline again to DM 27 billion in 1991.

Against the background of the drop in stock prices and the turbulence in the foreign exchanges in the latter part of 1987, the Federal Government decided in December 1987 to take fiscal measures to support economic growth. The main elements of this package were (1) DM 15 billion in additional loans to municipalities over 1988–90, with interest subsidies from the federal budget of DM 200 million annually over a ten-year period (at a total estimated cost of about DM 2.6 billion until the year 2000);[77] (2) DM 6 billion in additional loans under the Government's program for small and medium-sized enterprises;[78] and (3) an increase in the investment of the Federal Post System by DM 1½ billion to almost DM 20 billion in 1988. In addition, the Government announced that it would not take measures to counter a possible widening of the 1988 federal budget deficit and urged the lower levels of government to act similarly.

As early as January 1988, it had become apparent that there would be a significant revenue shortfall in the 1988 federal budget. The decline of the U.S. dollar at the end of 1987 to historically low levels vis-à-vis the deutsche mark necessitated a large write-off by the Bundesbank on its U.S. dollar reserves; this virtually eliminated profit transfers to the Federal Government, originally budgeted at DM 6 billion.[79] In addition, resolution of the budgetary problems of the EC required an increase of about DM 4 billion in Germany's contribution in 1988. As a result, the federal budget deficit was expected to increase by about DM 10 billion to about DM 40 billion, equivalent to 2 percent of GNP. The Government confirmed its determination not to offset the increase in the 1988 deficit but announced that it would take measures—specifically, an increase in certain consumption taxes—with the objective of reducing the federal deficit by some DM 10 billion in 1989.

Revenue estimates made in May did not change significantly the overall fiscal outlook for 1988 from the preliminary projections of January 1988. Compared with 1987, tax revenues of the Federal Government

---

[74] At the end of 1987, almost half a million people were enrolled in training programs or benefited from employment creation measures—11 percent more than at the end of 1986.

[75] Owing to the increase in the deficit of the territorial authorities, public debt increased from 41 percent of GNP in 1986 to 42 percent of GNP in 1987 (Table A31).

[76] A discussion of tax reform is included in Section IV.

[77] This allows a reduction of interest rates on these loans to about 2 percentage points below market rates. Since the local governments have to provide matching funds, this program is expected to support DM 40 billion of investment over the three-year period. It is not clear, however, to what extent this will be additional investment and to what extent it will simply be a change in the financing of investment that would have occurred anyway.

[78] This program is administered by Kreditanstalt für Wiederaufbau (KfW), a publicly owned bank. Under the program, KfW raises money in the capital market and extends loans to small and medium-sized enterprises at the prime rate, that is, without the interest markup customarily charged by commercial banks for credits to these enterprises.

[79] In April 1988, DM 240 million was transferred from the Bundesbank to the Federal Government, down from DM 7.3 billion in 1987 and DM 12.7 billion in 1986.

were expected to rise by ¼ of 1 percent, and, with the sale of two major government holdings in industry in early 1988 (Volkswagenwerk AG and VIAG, a holding company), capital revenues were projected to increase by 7 percent; however, a 31 percent drop in other revenues (chiefly Bundesbank profits) was projected to reduce total federal revenue to a level some 2 percent below that in 1987. Total expenditures, which were expected to increase by 2½ percent, were likely to remain within budgetary limits, but some risks were thought to exist in the area of subsidies. Support to coal mining and the commercial aircraft industry in 1988 had been estimated on the basis of assumed exchange rates of DM 1.8 and DM 2.0 per U.S. dollar, respectively. Thus, if the dollar remained below these rates vis-à-vis the deutsche mark on average in 1988, additional subsidies to coal would have to be paid and further support for the aircraft industry might have to be negotiated.[80]

For the states, only a small increase in the deficit was envisaged for 1988 as the slowdown in the growth of revenues was expected to be met with expenditure restraint. However, a larger increase in the deficits of the municipalities was foreseen, despite a somewhat smaller increase in expenditures than in 1987. This reflected static tax revenues owing to the tax cuts that took effect in 1988, and only a small increase in transfers from the Federal Government and the Länder.[81]

For the territorial authorities as a whole, revenue was expected to increase by only 1¼ percent in 1988 after 2¼ percent in 1987, reflecting tax cuts, sharply lower Bundesbank profits, and a decline in capital revenue as the privatization program of the Federal Government approached completion and privatization by states and municipalities proceeded at a slower pace. Despite projected increases of expenditures by the federal, state, and local governments of 3 percent or less, total expenditures were expected to increase by 3¼ percent, owing to larger transfers to the EC. Thus, the territorial authorities altogether were expected to exceed the spending growth target for the fourth year in a row, despite the projected adherence of the three levels of government to the expenditure growth target set by the Financial Planning Council. The deficit was expected to rise by about DM 13½ billion to DM 64½ billion (3.1 percent of GNP).

The surplus of the social security system was projected to fall to DM 4 billion (0.2 percent of GNP), owing to a smaller surplus in the pension fund[82] and a sharply higher deficit in the unemployment insurance system. The latter was due to the extension of unemployment benefits decided upon in 1987 and the shift of programs for the retraining and education of the unemployed from the federal budget to the accounts of the unemployment insurance fund. The expected deficit would not only leave the unemployment insurance system without any cash reserves by the end of 1988, but would also require a DM 1.1 billion transfer from the Federal Government. Thus, reflecting the increase in the deficit of the territorial authorities and the decline in the surplus of the social security system, the deficit of the general government was projected to increase by around DM 15 billion to DM 49 billion (2.3 percent of GNP) in 1988. The 4 percent projected increase in total expenditures was expected to leave the share of public expenditures in GNP more or less unchanged.

Stronger-than-expected economic growth in the course of 1988 contributed to higher tax revenue and thus reduced public deficits from the levels expected earlier in the year. This favorable development has also improved the medium-term outlook for public finances (see below).

Federal Budget for 1989

The draft federal budget for 1989 incorporates a 4.6 percent increase in expenditures; this is well above the 2½ percent increase envisaged in the Government's medium-term financial plan for 1989–91. Higher-than-expected outlays are due chiefly to transfers for unemployment insurance to the Federal Labor Office (DM 3.3 billion) and to transfers to "structurally weak" federal states (DM 2.5 billion). On the revenue side, in addition to the withholding tax on interest income (see Section IV) the draft budget includes a DM 8.1 billion increase in excise taxes. Of this, DM 6.6 billion are accounted for by higher taxes on gasoline and heating oil (DM 5 billion) and the introduction of a tax on natural gas (DM 1.6 billion). Additional taxes on tobacco and on insurance premiums are expected to yield DM 0.4 billion and DM 1.1 billion, respectively. Reflecting these tax increases, total tax revenues are expected to increase by 7 percent in 1989. Thus, with profit transfers from the Bundesbank budgeted at DM 5 billion, the federal deficit is expected to decline in 1989 to DM 32.9 billion (1½ percent of GNP).

---

[80] See also Section IV.

[81] A dampening effect on public expenditure growth was expected from the relatively small wage increase (of 2 percent for 1988 as a whole) for public sector employees. In addition, only a slight increase in capital expenditures was expected owing to sluggish investment spending by municipalities. Investment expenditures of local authorities would have declined in 1988 if the Federal Government had not decided to offer DM 5 billion in loans in this year at subsidized interest rates.

[82] Despite an increase in pensions of only 3 percent, a rise in the number of retirees and the gradual extension of pension benefits to all mothers born in the years before 1921 meant that pension outlays were likely to increase at a higher rate than contributions.

Medium-Term Outlook

Since 1986, when the emphasis in fiscal policy began to shift and the relative decline in social transfers ended, the medium-term outlook for government finances has changed considerably. This is illustrated in Table 5, which compares the staff's most recent fiscal projections with those prepared in 1986.[83]

The Government's earlier stress on quantitative consolidation partly reflected its conviction that the reduction of public expenditures and deficits would crowd in private investment and boost economic growth. But this has not occurred to the extent expected. At the same time, growing external surpluses have given rise to pressures from abroad to moderate the program of fiscal restraint. In shifting to tax reduction and reform despite a temporary setback in the quantitative consolidation of government finances, the Government has taken into account these concerns. Also, the view appears to have gained strength that too strong a reliance on the quantitative consolidation of government finances alone is not conducive to the revival of private economic activity, in particular investment. For this to happen, far-reaching structural reforms need to be implemented.[84] However, the Government's decision to accept a renewed rise in deficits and public indebtedness in order to make faster progress with tax reform, the main element of "qualitative" consolidation, is not without political and economic risks. From a political point of view, the Government runs the risk of being accused of repeating the sins of the past; from an economic point of view, higher public indebtedness may, in a longer perspective, be cause for concern in view of the pronounced demographic changes in prospect for the next two or three decades (see Section IV).[85]

Evaluation

The review of fiscal developments since 1982 raises the question as to whether the earlier progress and later slippage in quantitative consolidation was the result of deliberate fiscal policy decisions or of special factors and cyclical developments.

---

[83] The staff's recent projections were made in December 1988. They differ, therefore, from 1988 budgetary estimates contained in Tables A24–A28 and A30, which are the projections of the German authorities, based on revenue estimates made in May 1988.

[84] See Federal Republic of Germany (1988, pp. 12–19); Biedenkopf (1988); and Section IV.

[85] The deficit projected for the territorial authorities by 1992 is, however, still slightly above a level consistent with a stable debt-to-GNP ratio and nominal GNP growth of 4½ percent a year in the medium term.

Table 5. Medium-Term Fiscal Projections for the General Government

(In percent of GNP)

	1988	1989	1990	1991	1992
Budget balances					
1986 projections	−0.3	0.2	0.3	0.3	0.5
1988 projections[1]	−2.0	−0.8	−1.4	−1.2	−1.1
Public debt[2]					
1986 projections	41.0	40.3	39.6	38.9	38.0
1988 projections[1]	42.5	42.7	42.8	42.9	42.9

Source: Fund staff estimates.
[1] Made in December 1988.
[2] Gross debt of the territorial authorities.

The decline in budget deficits during the first half of the 1980s coincided with an increase in profit transfers from the Bundesbank to the Federal Government. The argument has therefore been made that the success in quantitative consolidation during this period and its later partial reversal had more to do with fluctuations in Bundesbank profits than with deliberate policy action. Table 6 gives the deficits of the federal and general governments excluding profit transfers from the Bundesbank.[86]

Table 6. Fiscal Balances Excluding Profit Transfers from the Bundesbank

(In percent of GNP)

	1981	1982	1983	1984	1985	1986	1987
Federal Government	−2.6	−3.0	−2.6	−2.3	−2.0	−1.9	−1.8
General government	−3.8	−3.9	−3.2	−2.5	−1.8	−1.9	−2.1

Source: Fund staff calculations.

Clearly, the reductions in fiscal deficits in 1981–85 were achieved independently of the increase in Bundesbank profits. Furthermore, federal deficits would have continued to decline in 1986–87, although at a much slower pace than in the years before, had there not been the fall in profit transfers from the Bundesbank. At the general government level, however, the process of quantitative consolidation would have been reversed in 1986–87 even if Bundesbank profits had remained at the level of previous years. This reflected the rapid rise in the deficits of the states and munici-

---

[86] This is not to imply that public financial balances should in principle be measured excluding net operating profits of the central bank (see Blejer and Chu (1988)). In Germany, however, accounting principles of the Bundesbank contribute to the year-to-year variation of Bundesbank profits.

palities as well as the decline in the surplus of the social security system during this period.

Could cyclical influences have distorted the picture? In order to answer this question it is necessary to look at the development of structural budgets, that is, those fiscal balances that would have prevailed if the economy had grown along its "normal" trend instead of going through an expansion with occasional growth recessions. The estimation of a structural budget balance begins with the computation of a "neutral" fiscal balance.[87] The ratios of government revenue and government expenditures (net of unemployment insurance benefits) to GNP in a year when the economy is operating at full capacity can be used for the calculation of neutral revenues and expenditures in the current year. Neutral revenues are calculated by applying the base-year revenue ratio to current GNP; neutral expenditures are derived from the base-year expenditure ratio and nominal potential output. The resulting neutral budget balance is then defined as the difference between the neutral revenues and neutral expenditures. The structural balance in any given year can then be approximated as the sum of the budget balance in the base year when the economy was at its potential and the cumulated differences between the actual and neutral balances since then. Table 7 below contains the results of these computations with 1978 assumed as the base year.

Table 7. Structural Budget Balances for the Federal and General Governments

(In percent of GNP)

	1981	1982	1983	1984	1985	1986	1987
Federal Government	−1.8	−1.1	−0.9	−1.1	−0.8	−0.8	−0.8
General government	−2.5	−0.6	−0.2	−0.8	0.1	−0.1	−0.2

Source: Fund staff calculations.

These results are subject to a number of simplifying assumptions and should therefore be interpreted with caution. For instance, results depend to a significant degree on the choice of the base year and the estimates of productive potential.[88] Nevertheless, when coupled with the earlier review of fiscal developments and policy decisions, these estimates can be taken as broad indicators of the underlying fiscal policy stance in Germany through the 1980s.

The estimated structural budget balances confirm the earlier historical analysis. Discretionary fiscal policy was instrumental in reducing public deficits in 1982–85[89] but the fiscal policy stance seems to have shifted somewhat in the years after 1985. Progress in quantitative consolidation appears to have ended at the federal government level (in 1986–87); but at the general government level there was a partial reversal owing to both buoyant expenditure growth of states and municipalities and the decline in the surplus of the social security system.

Policy Issues

The discussion of the medium-term economic prospects at the end of Section II highlighted the need for policy action to promote growth that was better balanced in terms of the labor market and the balance of payments. Much debate has focused on the extent to which use of conventional tools of aggregate demand management (i.e., fiscal and monetary policies) can contribute to this end. In light of the developments discussed above, the debate has resulted in a growing awareness that there is currently little room for maneuver in the use of these policy instruments.

Monetary policy has been fairly expansionary since the end of 1985 and, given the current relatively liquid state of the economy, there are concerns that the economy is now quite vulnerable to exogenous price shocks initiating a new inflationary spiral. Moreover, the hard-won credibility of the Bundesbank could be threatened by a continued expansion of money in excess of target; a loss of credibility would compromise the Bundesbank's ability to resist any future inflationary pressures. Furthermore, there are grounds for doubting the efficacy of monetary policy, or financial instruments, for achieving real goals such as durable employment and growth. Notably, over the past few years, the Bundesbank's efforts to lower short-term interest rates have had little impact on long-term rates, which are perhaps more important for the interest sensitive components of demand.

Fiscal policy is also constrained. As discussed in the previous section, the debt of the territorial authorities is expected to rise to about 43 percent of GNP by 1992, and there have been signs of unease both in political circles and in the financial markets.

---

[87] For a review of fiscal impulse measures see Heller and others (1986).

[88] See, for example, Thormählen and Leibinger (1987) for a critical discussion of recent estimates of the structural budget balance for Germany. The above estimates are also distorted by year-to-year variations of Bundesbank profits, related to the accounting practices of the Bundesbank.

[89] It is noteworthy that a considerable step toward quantitative consolidation was already taken in 1982, the year before the present conservative-liberal coalition was able to shape fiscal policy. This is not apparent from the actual budget balances in 1981–82, which changed only little during this period, since the economic downturn inflated the deficits of 1982.

Thus, there are limits to the use of monetary and fiscal policy to stimulate aggregate demand. Moreover, there are doubts about the desirability of a pronounced demand stimulus at this stage of the upswing; in spite of the high unemployment rate, the output gap in the German economy appears to be relatively small.[90]

Thus, it is not clear that aggregate demand stimulus, on its own, could significantly reduce the current account and labor market imbalances without generating inflationary pressure that would ultimately undermine any short-term gains. Indeed, there is a growing consensus that the fundamental economic problems in Germany are related to the supply side of the economy. In the following section, these supply side issues are considered.

---

[90] The output gap is the gap between potential output and actual output. In Section IV, evidence is presented that suggests that the high unemployment rate in Germany mostly reflects institutional rigidities and inappropriate relative factor prices rather than deficient demand.

IV Structural Policies

During the 1950s and most of the 1960s, the German economy performed well. A stable social and political climate as well as a favorable external environment contributed to the economic success of this period. In the early 1970s, however, major changes in the external environment took place: the Bretton Woods system of fixed exchange rates was replaced by a floating rate system, mineral oil prices increased sharply, and other industrial countries, in particular Japan, emerged as major players in the arena of international trade. In addition, the domestic social fabric changed. With the retirement of the postwar generation of industrial and trade union leaders, the social consensus between trade unions and employers weakened. Also, the profound demographic changes of the late 1950s and mid-1960s—specifically the initial increase and later fall in birth rates—began to influence the educational system and the labor market. As a result of these developments, the German economy entered a period of substantial structural change.

So far, adjustment has been achieved only in part. Capacity in manufacturing has been expanding by less than would have been necessary to maintain employment in this sector; the services sector has not expanded fast enough to offset the decline in employment in manufacturing, agriculture, and other primary sectors and to absorb a growing labor force; and the decades-old problems of the ailing industries, for example, coal mining, iron and steel, and shipbuilding, have continued. In the judgment of the Council of Economic Advisors, the German economy has not been able to develop the flexibility and dynamism necessary for successful adjustment.[91] Consequently, economic growth has been lower on average and less stable during the last fifteen years than it was during the postwar cycle. The factors—labor, capital, and total factor productivity—that contributed to the slowdown are discussed in more detail below.

Over the last three years in particular, external pressures for adjustment have increased substantially. When efforts to redress the international economic imbalances among major industrial countries began in 1985, the emphasis was on aggregate demand policies and exchange rate changes. However, with the room for maneuver in these policy areas diminishing and the adjustment being far from complete, the focus has shifted to structural policies. For Germany in particular, the view has gained strength that trade liberalization, as well as deregulation of goods, services, and labor markets, would enhance the responsiveness of the economy to price signals (and especially recent exchange rate changes) and thus contribute to external adjustment. Yet economic policy in Germany has not always been conducive to structural change. In fact, policies on trade and industry have often had the effect of entrenching existing structures.

In its 1988 Annual Report, the Government has reaffirmed the need for structural policy action and the phasing out of subsidies that inhibit structural adjustment.[92] So far, the Government has completed plans for the 1990 tax reform (which includes a reduction of tax exemptions), made proposals for a reform of the medical and pension insurance systems, appointed a commission of experts to study areas of possible deregulation, proposed a modification of shop opening hours, and introduced legislation with a view to increasing competition in the telecommunications sector. But there is little doubt that more needs to be done.

A discussion of structural policy options for Germany essentially needs to address two basic questions: (1) What are the areas where policy action is needed? and (2) What effects can be expected from structural measures? Neither question is easy to answer; consequently, a first tentative essay at some answers is given rather than a blueprint for structural policy. The most important developments and prospects in the areas of tax reform, social policy, trade and industrial policy, and privatization and deregulation are reviewed. The purpose is to give background information about existing institutional arrangements that may

---

[91] Sachverständigenrat zur Begutachtung der gesamtwirtschaftlichen Entwicklung (1987).

[92] See Federal Republic of Germany (1988, pp. 14–15).

hinder structural change and to identify areas where policy action may be appropriate. The basic message from this discussion is that the authorities' present plans for reforms are important steps in the right direction but that more comprehensive and wider-ranging reform measures are still needed.

The review of the institutional setting and policy developments is then complemented by a quantitative analysis of the effects of structural reform on economic growth in Germany in a forward-looking framework. This poses a few methodological problems. Projections of economic developments under alternative structural policy measures require a model framework that is capable of both projecting macroeconomic aggregates and tracing the microeconomic effects of policy-induced relative price changes. No such model exists for Germany: the dynamic macroeconomic models are incapable of dealing with structural change while the models that are capable of dealing fully with relative price effects at a disaggregated level are comparative static models. The exercise conducted in the last part of this section therefore combines a dynamic macroeconomic model with a comparative-static, multisector, microeconomic model of the German economy.

The simulation results show that reducing subsidy payments and removing trade restrictions in agriculture, coal mining, and the iron, steel, shipbuilding, textiles, and clothing industries would raise GNP growth and employment and lower inflation and the external trade surplus in the medium term. Trade liberalization and subsidy reduction would not only improve the medium-term growth prospects, however, but would also increase the flexibility of the German economy in responding to the recent sharp appreciation of the deutsche mark. In a separate but related study (Mayer (1988)), it has been argued that subsidies and other nontariff barriers to trade in certain key sectors—agriculture, coal mining, iron and steel, shipbuilding, textiles and clothing—insulate these sectors and industries from the effects of relative price changes, in particular, exchange rate changes. As a consequence, the burden of adjustment is shifted from these relatively inefficient sectors and industries to unprotected and efficient sectors. Moreover, labor market rigidities and numerous regulations affecting in particular the services sector impede the reallocation of resources and diminish the supply response of the nontraded goods' sector to improvements in the terms of trade associated with a real appreciation of the deutsche mark. As a result of protection, rigidities, and regulations, the contractionary effects of an exchange rate appreciation tend to dominate the expansionary effects with negative consequences for economic growth and employment.

The review and analysis of structural policy in this section prompts an obvious question: if areas for reform can be clearly identified and the economic effects of these reforms are as positive as suggested here, why is it then that progress has been so slow and limited? To answer this question one must venture into political economy. The benefits of structural reform are macroeconomic, but within the macroeconomy there will be those that gain and those that lose from any structural change. In many cases, the gainers are large and heterogeneous groups in society, such as consumers or the labor force, while the losers are small and homogeneous groups, such as farmers or coal miners. The interests of the gainers are difficult to organize while the losers are usually capable of forming very effective political lobbying groups (Olson (1971)). In a political system where the swing voters often decide election results, politicians are wary of antagonizing well-organized social groups.[93] Bold structural reforms are thus perceived as politically impossible unless there is a real crisis that welds together public opinion. In contrast, the gainers and losers of macroeconomic policies are generally heterogeneous groups, which are less capable of mounting fierce lobbying efforts. This explains, in part, politicians' preference for macroeconomic instruments in achieving certain economic objectives even when the problem is structural in nature.

The Slowdown in Economic Growth

In order to examine the main causes for the slowdown in economic growth that has taken place during the last fifteen years or so it is useful to decompose the economy's production function into its three principal components—labor, capital, and total factor productivity.

Labor

One reason for the unsatisfactory growth performance of the German economy can be seen in the labor market. With the demand for labor significantly lagging behind supply, the rate of unemployment increased from less than 1 percent of the labor force in 1970 to 8 percent in 1987. Most of the studies of

---

[93] This is hardly a new discovery. For example, in 1787 James Madison observed: "A landed interest, a manufacturing interest, a mercantile interest, a moneyed interest, with many lesser interests, grow up of necessity in civilized nations, and divide them into different classes, actuated by different sentiments and views. The regulation of these various and interfering interests forms the principal task of modern legislation and involves the spirit of party and faction in the necessary and ordinary operations of government" (Madison, 1961, p. 79).

this problem have concluded that much of the rise in unemployment in Germany is attributable to structural phenomena, as suggested by a rise in the hypothetical rate of unemployment that would prevail under non-accelerating inflation (the NAIRU, or, a related concept, the natural rate of unemployment). The studies of Layard and Basevi (1984), Coe (1985), Coe and Cagliardi (1985), and Burda and Sachs (1987) lead to the conclusion that the NAIRU has increased from between ½ of 1 percent and 1½ percent in 1970 to up to 8 percent by the mid-1980s, roughly in line with the rise in actual unemployment.

Burda and Sachs examined to what extent the increase in the NAIRU could be accounted for by a rise in frictional unemployment; such a rise in frictional unemployment could have been due either to a greater amount of time spent looking for a new job in response to reduced costs of unemployment or it could have been due to greater mismatch (occupational, industrial, or regional) between available jobs and jobseekers. They discounted the possibility that the cost of job search has been reduced on the grounds that there has not been an increase in the generosity of unemployment benefits. Nor did they find evidence of increased mismatch in either an occupational or industrial classification of jobs and jobseekers; they did, however, find some evidence of increased mismatch between regions. The view that there has not been a significant rise in frictional unemployment is supported by the fact that there does not seem to have been a pronounced shift in the so-called Beveridge curve, which maps the relationship between vacancies and unemployment (Chart 12). Since the mid-1970s the German labor market has moved along the Beveridge curve to a situation of high unemployment combined with a low vacancy rate.

Thus, it appears that the rise in the NAIRU has been due to insufficiently flexible real wages, as a result of which real wages have not adjusted to levels that would allow a higher degree of utilization of the labor force. Artus (1984), Lipschitz (1986), and Burda and Sachs (1987), all find that substantial wage gaps—defined as the difference between actual real wages and the level of real wages that would be required to ensure a desired level of employment—emerged in the German manufacturing sector in the 1970s, and the latter two studies indicate that these large wage gaps persisted in the first half of the 1980s.

Supporting these studies, a plot of the Okun curve (which traces the relationship between unemployment rates and capacity utilization) shows the development of a perverse relationship—that is, the concurrence of high levels of capacity utilization with high rates of unemployment (Chart 13). This would seem to suggest that, at current relative factor prices, there is little

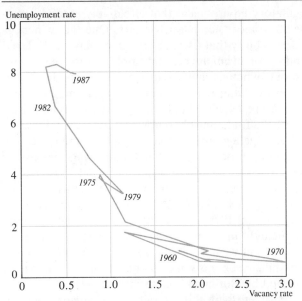

Chart 12. Federal Republic of Germany: The Beveridge Curve

(In percent of labor force)

Sources: Statistisches Bundesamt, *Volkswirtschaftliche Gesamtrechnungen* and *Statistisches Jahrbuch*, various issues.

scope for absorbing additional labor. While these two pieces of evidence relate to the manufacturing sector, the problem appears to be more generally applicable to the economy as a whole.

The inadequate responsiveness of real wages to labor market conditions can be attributed in significant

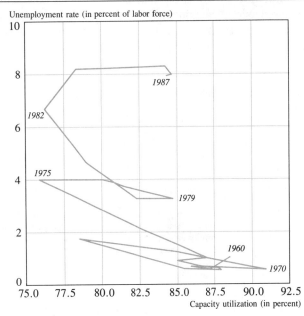

Chart 13. Federal Republic of Germany: The Okun Curve

Sources: Statistisches Bundesamt, *Volkswirtschaftliche Gesamtrechnungen*, various issues; and Ifo Institute, *IFO Schnelldienst*, various issues.

part to the nature of the collective bargaining process in Germany. The coverage of collective bargaining agreements is very wide; this results not so much from the degree of unionization[94] but from factors in the regulatory environment that effectively extend agreements to most nonunion workers. One factor here is the possibility that a wage settlement may be declared binding for all employees and employers in the relevant sectors, whether unionized or not.[95] Second, all employers holding membership in a bargaining confederation are legally bound by agreements reached on their behalf, and membership rates are generally high.

The wage agreements that result from the collective bargaining process specify minimum wage levels for the various steps of the wage scale and thus restrict the responsiveness of labor costs to demand and supply conditions. Gundlach (1986) provides evidence that, after the mid-1970s, these lower wage bounds became significantly more binding and also reports a conspicuous lack of wage flexibility between sectors, regions, and skills over the same period. Similar evidence is reported by Burda and Sachs (1987).

Consistent with this overall picture, wage developments in nontradables sectors in Germany have increasingly paralleled those in manufacturing. This reflects, inter alia, the fact that the degree of unionization in important nontradables sectors—banking, insurance, food and restaurants, for example—has risen considerably since the early 1970s; there has also been a significant number of wage contract extensions (under the wage contract law) in the services sector.[96] As the tradables sectors were hurt by various supply shocks in the 1970s, insufficiently flexible real wages resulted in a shedding of labor from these sectors. On each occasion, the nontradables sectors have not been able to absorb these workers, as these sectors also suffered from relatively inflexible real wages (Chart 5).

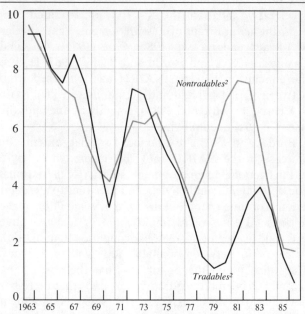

Chart 14. Federal Republic of Germany: Capital/Labor Ratios

(Annual change in percent)[1]

Source: Staff calculations based on Statistisches Bundesamt, *Volkswirtschaftliche Gesamtrechnungen*, various issues.
[1] Three-year moving average over the current and the previous two years.
[2] Tradables comprise manufacturing, agriculture, and transport and communication. Nontradables comprise public and private services, construction, and trade. The energy sector is not included.

Consistent with the high wage growth in the nontradable sectors, there have been sharp increases in capital/labor ratios in these sectors (Chart 14). Consequently, in comparison with other countries, Germany has seen a relatively rapid pace of productivity growth in the nontradables sectors but also slower employment growth. Thus Ochel and Wegner (1987) report that, from 1973 to 1984, labor productivity in private services in Germany increased at an annual average rate of 2.6 percent compared with 1.5 percent in France, 1 percent in the United Kingdom, 0.3 percent in the United States, and 0.2 percent in Italy. At the same time employment in private services rose at a much lower rate in Germany (0.4 percent) than in France (1.7 percent), the United Kingdom (1.2 percent), the United States (2.7 percent), and Italy (2.6 percent).[97]

A number of other factors have influenced labor market developments in Germany. First, the wide coverage, high replacement ratio (i.e., the ratio of unemployment benefits to wages) and the long duration of unemployment benefits in Germany provide unemployed workers with greater freedom to wait out

---

[94] In 1986, about 45 percent of the employed wage and salary earners subject to social security taxes were members of a union (Statistisches Bundesamt, *Statistisches Jahrbuch, 1987*, Tables 6.8.1 and 26.13).

[95] According to the so-called *Allgemeinverbindlichkeitserklärung*, Section 5 of the Wage Contract Law (*Tarifvertragsgesetz*) of 1969, a State Minister of Labor may, at the request of either bargaining party, declare an agreement generally binding if this is deemed in the public interest, provided more than half the employees in the relevant sector are employed by firms party to the agreement. Such a declaration renders an agreement binding for all employees irrespective of union membership and all employers irrespective of membership of an employer's federation. While only a relatively small number of wage agreements are subject to a declaration in any year, the possibility that a request may be made to a Minister of Labor influences the behavior of unions and employers' associations.

[96] In many service industries employer membership in bargaining confederations is more limited than in manufacturing, and thus fewer employers are legally bound by agreements unless these agreements are extended.

[97] The comparison between developments in the services sector in the United States and Germany is particularly striking and is discussed in detail in Burda and Sachs (1987).

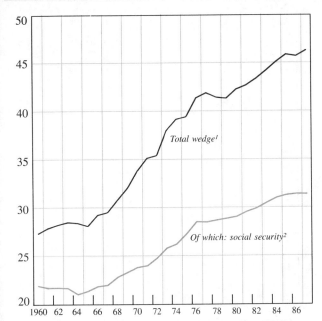

Chart 15. Federal Republic of Germany: The Wedge Between the Net Consumption Wage and the Gross Product Wage
(In percent)

Source: Statistisches Bundesamt, *Volkswirtschaftliche Gesamtrechnungen*, various issues.
[1] Income tax plus employers' and employees' social security contributions as a fraction of total compensation of employees. This is equal to unity minus the net labor income as a fraction of the gross labor costs.
[2] Employers' and employees' social security contributions as a fraction of total compensation of employees.

periods of unemployment than in some other countries.[98] Second, the so-called social plans raise firing costs by requiring employers to make substantial payments to employees at times of large layoffs. Third, the wedge between the gross product wage and the net consumption wage has grown over time (Chart 15); the impact of this in an environment of real consumption wage rigidity is to reduce employment in the those sectors that are exposed to foreign competition while at the same time impairing the ability of the nontradables sectors to absorb the labor shed elsewhere (see Burda (1988)). Finally, a lack of flexibility in the housing market is also said to have affected the geographical mobility of labor.

On the positive side, action has been taken to improve the legal conditions governing the hiring of youth and part-time workers,[99] and more flexible working hours are being adopted in some industries.

Capital

Although institutional factors in the labor market provide many possible sources of the rise in unemployment and the slowdown of growth in Germany, the possibility of a capital shortage has also received considerable attention. This is perhaps a natural question arising from Chart 13; if capacity utilization is very high at current unemployment levels, does not this suggest the need for additional capacity? A quick look at the data would also seem to lend some support to the proposition that many of the current difficulties can be traced to an insufficient level of capital formation. For example, Burda (1988) observes that since 1973 the gross capital stock in Germany has grown much more slowly relative to the pre-1973 trend than in any of the other six industrial countries for which he presents data.[100] The proposition that investment has been insufficient would also seem to be supported by the observation that the share of GNP devoted to fixed investment has been on a secular decline in Germany and that the rate of growth of investment during the current upswing has been lower than during previous upswings.

A closer examination of the historical evolution of fixed investment in Germany, however, reveals that, in an historical context, business investment has not been as weak recently as might appear to be the case. It is clear from Chart 16 that the ratio of investment to GNP in current prices has declined substantially since the 1960s and this holds true for all the main components of investment; however, when one looks at gross fixed investment in terms of constant 1980 prices, the picture is altered somewhat.[101] In particular, while real private investment has declined relative to GNP, this has been due principally to residential construction and to a lesser extent business construction; on balance, machinery and equipment investment has shown a slight upward trend and business investment, in total, has declined by a relatively small amount (Chart 17).[102] The different behavior of nominal and real investment relative to GNP reflects changing relative prices in the economy; the price of machinery and equipment has declined significantly relative to the aggregate value-added deflator while that of construction goods has risen.[103]

---

[98] The comparison with the United States is striking, see Burtless (1987).
[99] The "*Beschäftigungsförderungsgesetz*" of 1985.

[100] The other countries are Austria, France, Italy, Sweden, the United Kingdom, and the United States.
[101] From a supply-side perspective, real investment would appear to be a more appropriate concept than investment expressed in nominal prices.
[102] Real business gross fixed investment was about ½ of 1 percentage point of GNP lower on average during the period 1980–87, compared with the period 1960–69.
[103] Thus, the drop in the ratio of construction investment to GNP has been greater in constant prices than in current prices while the converse applies to machinery and equipment investment.

IV • STRUCTURAL POLICIES

Chart 16. Federal Republic of Germany: Nominal Gross Fixed Investment, 1960–87

(In percent of GNP at current prices)

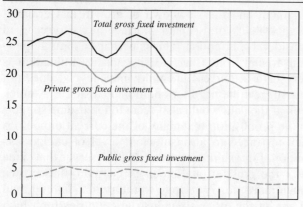

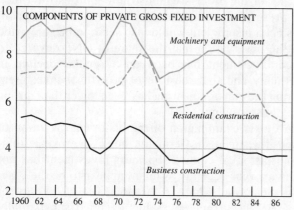

Source: Statistisches Bundesamt, *Volkswirtschaftliche Gesamtrechnungen*, various issues.

Chart 17. Federal Republic of Germany: The Components of Real Private Gross Fixed Investment, 1960–87

(In percent of GNP in 1980 prices)

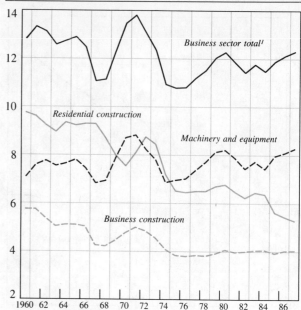

Source: Statistisches Bundesamt, *Volkswirtschaftliche Gesamtrechnungen*, various issues.
[1] Machinery and equipment plus business construction.

Chart 18. Federal Republic of Germany: Growth of Real Business Sector Net Capital Stock and Its Components, 1960–87

(In percent a year; in 1980 prices)

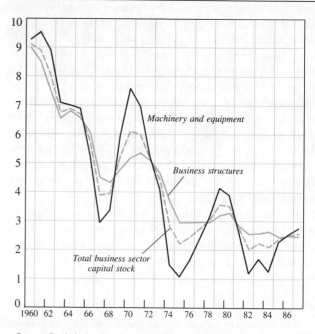

Source: Statistisches Bundesamt, *Volkswirtschaftliche Gesamtrechnungen*, various issues.

While gross business investment does not seem particularly weak by historical standards, the share of depreciation of business capital in GNP has risen, reflecting the increase that has occurred in the capital intensity of production.[104] Thus, the ratio of net business sector investment to GNP has fallen significantly and this has been mirrored in a sharp decline over time in the rate of growth of the business sector capital stock (Chart 18).

These historical developments need to be seen in the context of changes in relative factor prices that have occurred since the early 1960s. The price of capital services has declined sharply relative to other factor services, owing principally to a decline in the relative price of capital goods and, in particular, of machinery and equipment.[105] This change in the relative prices of capital goods encouraged a sharp increase

---

[104] By capital intensity of production is meant the ratio of the net capital stock to GNP, both measured in constant 1980 prices.

[105] The price of capital services includes financing costs (the cost of debt and equity finance) and tax factors, as well as the price of capital goods.

Chart 19. Federal Republic of Germany: Capital Intensity of Production, Relative Factor Prices, and the Return to Capital

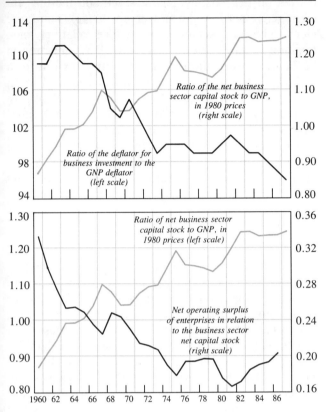

Source: Statistisches Bundesamt, *Volkswirtschaftliche Gesamtrechnungen*, various issues.

in the capital intensity of production. Chart 19 illustrates the close negative correlation between the capital intensity of production and the price of capital goods relative to the GNP deflator. It is seen that the changes in relative factor prices and the associated capital deepening were concentrated in the period prior to the mid-1970s; as relative price developments over this period favored a large increase in the capital intensity of production, net investment had to be sufficiently high to realize the transition to higher capital/output ratios. With the less pronounced changes in relative prices between 1974 and 1985, the incentives for increased capital intensity of production diminished. Thus, a downward trend in the share of net investment in GNP is not particularly surprising; it has been an inevitable consequence of the earlier boost to net investment resulting from the changing pattern of relative prices in the economy. Whether the rapid decline in the relative price of capital services through the mid-1970s was entirely appropriate is another matter.[106] It is also worth noting that the sharp decline in the return to capital, which has been a cause of concern to many, needs to be seen in the context of the significant increase in the capital intensity of production.[107]

How then do we evaluate the decline in investment as a source of the slowdown of economic growth in Germany? It is useful to consider what economic theory tells us about the optimal level of the capital stock. Using a conventional constant elasticity of substitution (CES) production function, it can be shown that when technical progress is labor augmenting, the equilibrium capital/output ratio depends only on the price of capital services relative to the price of output; thus with a constant relative price of capital services and labor augmenting technical progress, one would expect a constant equilibrium capital/output ratio.[108] The recent path of the capital/output ratio does not seem out of line with relative factor prices: with the price of capital goods falling relatively little in relation to the GNP deflator between 1973 and 1985, there seems little reason for there to have been a more pronounced increase in the capital intensity of production over this period. Indeed, it is seen in McDonald (1988a) that a theoretical framework of the type briefly outlined above explains the evolution of investment in machinery and equipment in Germany quite well for the period since 1960.

While changes in relative factor prices may explain quite well the evolution of the capital/output ratio and thus the share of investment in GNP, the growth of the capital stock has also inevitably been affected by developments relating to the two other principal contributors to output growth, that is, the labor supply and total factor productivity. Total factor productivity, in particular, has slowed considerably since the mid-1970s (see pp. 36–37); the effect of this has been to slow output growth and consequently the growth of the capital stock.

Finally, the growth of business investment in Germany following the 1982 recession has been slower than that after previous recessions. In addition to the

---

[106] This reflected, inter alia, the growth of labor market rigidities and a sharp rise in real wage gaps that were dealt with at some length in Lipschitz (1986) and were also discussed in the previous section, dealing with the labor markets.

[107] Conventional production functions show that the rate of return to a factor falls with the intensity of its use in the production process.

[108] Labor-augmenting technical progress manifests itself in labor becoming more efficient over time. The empirical evidence in McDonald (1988a) suggests that technical progress in Germany has been principally labor-augmenting. This is also the type of technical progress assumed in Lipschitz (1986) and Burda and Sachs (1987). Only labor-augmenting technical progress is consistent with a steady state. With capital-augmenting or disembodied technical progress, a constant price of capital relative to the price of output would produce a falling capital/output ratio. Autonomous technical progress comes through total factor productivity and has no influence on the equilibrium capital/labor ratio.

developments in factor prices and total factor productivity, which have already been discussed, it is worth noting that business investment in proportion to GNP fell more sharply in the recessions of 1967–68 and 1974–75 than in 1982; from this fact alone one would expect a less steep recovery of investment in the upswing.

That the slowdown of business investment in Germany since the mid-1970s does not appear to reflect any change in the underlying behavior of investors compared with earlier periods should not be taken to mean that the current level of the capital stock is optimal; if distortions in the economy have curtailed the rate of technical progress or have otherwise restricted the return to capital, then policy measures to alleviate those distortions are both desirable and likely to boost capital formation. Clearly, producers have little incentive to expand the capital stock at a faster rate without such policy measures, since all this would do is push the return to investors down further (by reducing the marginal product of capital or increasing the cost of capital).[109] Policy measures that boost the return to capital, however—such as those that reduce rigidities in the labor market, lower the taxation of capital income, or otherwise improve the allocation of capital so as to increase the dynamism of the economy—would increase the demand for capital and thereby boost net investment. The econometric analysis of machinery and equipment investment in McDonald (1988a) suggests that investment would respond quite vigorously to such measures.[110]

The scope for policy measures is readily apparent. First, the foregoing discussion of the labor market indicates the need for an easing of institutional rigidities in this market, with a view to reducing the real cost of labor. This would induce employers to hire more labor at the current level of the capital stock thereby reducing unemployment and increasing output. In the process the rate of return on the capital stock would be increased, inducing employers to increase the capital stock and thereby boosting net investment (until the desired new level of capital stock was reached) as well as creating a demand for additional labor.[111]

Second, dismantling the sectoral policies which support an inefficient allocation of resources across sectors would also be expected to boost the rate of return earned on the existing capital stock and thereby encourage greater investment (see pp. 40–46). Third, the German business sector currently faces relatively high taxation by international standards.[112] A reduction in business taxes would increase the after tax return to capital relative to the cost of capital and lead to an increase in the desired stock of capital relative to output. Again this would boost net investment until the desired new capital/output ratio was reached. Finally, structural policy that increased the dynamism (the rate of technical progress) in the economy would result in increasing returns to capital through time and lead to a permanently higher rate of net investment.

Total Factor Productivity

The German economy, like most other industrial economies, has seen a marked reduction in the rate of technical progress since the mid-1970s.[113] The sources of this slowdown are difficult to pin down.[114] In part, it is a natural outcome of the catch-up process.[115] It seems quite likely, however, that the same rigidities that have given rise, for example, to the high unemployment rate have also undermined the dynamism of the economy. One can imagine a number of ways in which this could result. For example, consider the possibility that the benefits of technological progress are embodied in additions to the capital stock. Labor market rigidities, as explained earlier, have reduced the growth of capital stock and are thus likely to have slowed technical progress. Similarly, large government subsidies have resulted in excessive employment in inefficient and moribund industries and since such industries are unlikely to be heavy investors, the rate of technical progress is likely to have been adversely affected. More generally, it seems a reasonable proposition that excessive regulation and rigidities, whose effects in the German economy seem to have become more constraining since the mid-1970s, have stymied initiative in the development and application of new technology.[116] The structural policy options to counter

---

[109] In principle, assuming constant returns to scale, and given the significant unemployment rate, one could increase the capital stock and avoid a reduction in the marginal product of capital by absorbing sufficient labor to maintain the prevailing capital/labor ratio. This would only be possible, however, if unemployed labor was equal in quality with employed labor and if an increased supply of capital did not require a higher return to capital.

[110] The analysis in McDonald (1988a) indicates that the capital/output ratio desired by investors would respond with an elasticity of a half to changes in the real user cost of capital; this result is consistent with the results obtained by Lipschitz (1986) in his study of the German manufacturing sector.

[111] Under this scenario, the capital/output and investment/output ratios would decline, but the effects of this on investment would be more than offset by the impact of a higher output level.

[112] See Leibfritz and Parsche (1988).

[113] See OECD (1988) and Englander and Mittelstädt (1988).

[114] See Englander and Mittelstädt (1988).

[115] Countries that were technologically behind in the early postwar years experienced very rapid technical progress during the catch-up period.

[116] As noted earlier, Gundlach (1986) provides evidence that the adverse implications of the collective bargaining process for labor market flexibility appear to have become more prominent since the mid-1970s. Moreover, there has been a pronounced shift in the composition of government subsidies away from general subsidies to subsidies for weak sectors, such as mining, iron and steel, and shipbuilding.

the slowdown of economic growth in Germany are discussed below.

Tax Reform

In their programmatic paper of December 1985, the Federal Ministry of Finance described the role of fiscal policy in the broader context of structural policy as follows:

> In an economic system based on free enterprise, economic growth is not so much the aim as the result of market processes. The task of the public sector is not to realize the highest possible growth rates at the cost of unwarrantable fiscal policy measures, but to ensure that economic activity can develop unhindered and is provided with sufficient incentive. The price signals transmitted by the market must reach the recipients . . . with as little distortion as possible.[117]

Tax reform with a view to reducing the distortionary effects of the tax system has been an integral part of the economic program of this Government. Given the high tax burden in Germany (Table A32), tax reform has had to go along with tax reduction.

In the process of tax reform to date, there have been three major steps, each one associated with tax reduction. In 1986, the basic tax allowance and additional allowances for families with children were increased, and the marginal income tax rate schedule was slightly flattened. These measures amounted to a tax cut of DM 11 billion (0.6 percent of GNP). Two years later, in 1988, the marginal tax rate schedule was further flattened, basic and education allowances were increased, and depreciation allowances for small and medium-sized enterprises were raised. The tax relief provided by these measures amounted to about DM 14 billion (0.7 percent of GNP). The larger part of these tax cuts (DM 9 billion) had already been decided upon in December 1984, together with the 1986 tax reductions, while the remainder consisted of measures originally included in the 1990 tax reform package that were brought forward to 1988.[118]

The third step in the tax reform is to take effect in 1990 and will entail tax reductions of about DM 37 billion, offset in part by cuts in tax exemptions and a closing of tax loopholes that should raise revenues by about DM 18 billion (Table A33). The tax cuts reflect a comprehensive tax rate reform (DM 32 billion), relief to families (DM 2.2 billion), an increase in tax allow-

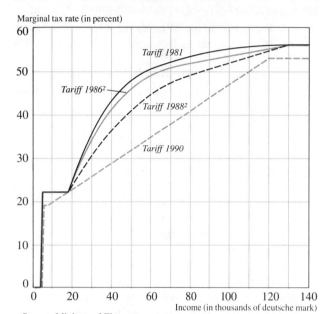

Chart 20. Federal Republic of Germany: Tax Reform and Marginal Income Tax Rates[1]

Source: Ministry of Finance.
[1] For an individual tax payer.
[2] Shown here only for incomes greater than DM 18,035.

ances for provisionary expenses (DM 0.6 billion), and a reduction of the corporation tax (DM 2.5 billion) (Chart 20). The reduction in tax exemptions and closing of tax loopholes consists of 62 measures that include changes in the area of income, corporation, trade, and other taxes as well as in special provisions and legal and administrative ordinances; in addition, some tax exemptions that expire in 1990 will not be renewed (Table A34).[119] About DM 7½ billion of the total package of DM 18 billion represents reductions of subsidies in the narrow definition of the Federal Government's Subsidy Report; if measures to widen the

---

[117] Federal Republic of Germany (1985, p. 19).
[118] The decision to bring forward DM 5 billion of the 1990 tax cuts to 1988 was announced on the occasion of a Group of Six meeting on February 22, 1987 (Louvre Accord). This, it was thought, would help bolster domestic demand and contribute both to stronger growth and a more rapid reduction of the current account surplus.

[119] The most important, and controversial, single measure is the introduction of a 10 percent withholding tax on interest income with effect from 1989. Interest income is in principle subject to income taxation; however, since interest payments are not reported to the tax authorities by financial institutions, the declaration of interest income has been left to the taxpayers. This has given rise to widespread tax evasion. Under the new law, the financial institutions will be responsible for collecting and transferring to the authorities the withholding tax on interest income. Interest payments to both domestic and foreign residents will be subject to this tax. The financial institutions will issue a certificate for the amount paid, which the taxpayer then may deduct from his income tax obligations when he files his income tax declaration. The same procedure will apply to foreign residents who are covered by a double taxation agreement of their home country with the Federal Republic of Germany. Interest on savings accounts and part of the interest accumulated in life insurance schemes will be exempt from the withholding tax. In addition, churches, charitable institutions, political parties, pension plans, professional associations, and certain banks engaged in the execution of public programs will be completely or partly exempted from the tax.

tax base are also counted as reductions in subsidies in the form of tax preferences, total subsidy cuts amount to about DM 12 billion. The rest represents measures to improve the enforcement of tax collection, such as the withholding of taxes on interest income at source or the charging of interest on outstanding tax liabilities.

The three steps of tax reform together with increased depreciation allowances on commercially utilized buildings introduced earlier (about DM 4 billion as of 1990) will provide a total (net) reduction in direct taxes of some DM 48 billion by 1990 (about 2¼ percent of projected GNP).[120] As a result, according to projections by the Ministry of Finance, the ratio of taxes of the territorial authorities to GNP will decline from almost 25 percent in 1980 to 22½ percent in 1990—the lowest level since the 1950s.[121] Furthermore, tax reform will increase, albeit slightly, the weight of indirect taxes relative to direct taxes thereby reversing the developments of the 1970s with respect to the tax structure (Table 8). Since direct taxes affect the creation of income, while indirect taxes have a greater effect on the use of income, a shift from the former to the latter is expected to strengthen incentives for work, investment, and, eventually, growth.

While the tax reform represents a major step toward a simpler, fairer, and less distortionary tax system that will have positive effects on employment and growth in Germany, it is worth noting that recent tax reforms in other countries have been more radical. This is illustrated in Table A35 which shows previous, present, and proposed top marginal tax rates in a number of countries. At present, the top German marginal tax rate excluding social security contributions is slightly below the average for the 24 mostly industrial countries.[122] The proposed top marginal tax rate, however, is the second highest of the sample of those countries with proposed tax reforms and well above the sample average.

Moreover, business taxation will continue to be high by international standards[123] and the stock turnover tax (*Börsenumsatzsteuer*) and the company tax (*Gesellschaftssteuer*), which hinder the development of

Table 8. Direct and Indirect Taxes in Percent of Total

	1960	1970	1980	1986	1988	1990
Direct taxes	54	54	59	59	59	58
Indirect taxes	46	46	41	41	41	42

Source: Ministry of Finance.

German financial markets and put joint stock companies at a disadvantage, will remain in force. The trade tax (*Gewerbesteuer*), which affects businesses differently according to size and location, is widely regarded as a major source of distortion; but it, too, will not be abolished as it is the main source of revenue for municipalities.[124] High business taxes, however, contribute to a low return on fixed capital investments in Germany relative to domestic financial investments or investment abroad and, as explained above, are therefore regarded as one of the obstacles to higher investment and growth in Germany.[125]

Reform of the Social Security System

During the 1970s, social security transfers grew much faster than nominal GNP. By 1982, social security transfers had reached almost 18 percent of GNP, up from 13 percent in 1968–73 (Table A36). This put Germany well above the average for major industrial countries and imposed a heavy burden on wage earners and employers, who finance the larger part of these expenditures with their social security contributions.[126] Since 1982, the year when the present coalition came to power, the increase in social security expenditures has slowed; indeed, in relation to GNP, social security spending has dropped by 1½ percentage points to a little under 16 percent in 1987. The reform measures implemented to date, which are discussed below, have been effective in solving the immediate problems of the German social security system; but, given the obstinacy of the unemployment problem and the prospective increases in the average age of the population over the next two to three decades, more needs to be done.

The Pension System

Reflecting the low birth rate since the mid-1960s, the average age of the German population will rise rapidly in the 1990s and thereafter. Table A37 contains

---

[120] Here, and elsewhere in the paper, direct taxes are defined to exclude social security contributions.

[121] Cuts in direct taxes, however, will be partly offset by increases in indirect taxes in order to finance higher contributions by Germany to the EC budget and transfers to the Federal Labor Office and "structurally weak" federal states (see Section III).

[122] Present rates including social security contributions, however, are well above the sample average.

[123] See Leibfritz and Parsche (1988). Their calculations indicate that under the present tax system, profits of a company producing machinery and equipment, for example, are taxed at an effective rate of about 55–60 percent in Germany, but at only 44 percent in the United States and 23 percent in Switzerland. Tax reform in Germany will not narrow this gap significantly.

[124] There are, however, plans for a continuation of the tax reform process during the next legislative period (1991–94) focusing in particular on a reform of business taxation.

[125] See pp. 30–37 and Engels (1988).

[126] About 18 percent of the expenditures of the pension system are covered by contributions from the federal budget.

estimates of the dependency ratio (defined as the ratio of persons aged 65 and older to those aged 15–64) for Germany and six other major industrial countries. Because of the baby boom of the 1950s and early 1960s, the dependency ratio in Germany will fall from 23 in 1980 to 21 in 1990. In that year, it will be not much above the average for the seven industrial countries and lower than in the United Kingdom and Italy. In the following decades, however, it will increase rapidly and is forecast to reach 33½ percent by 2020, well above the Group of Seven average and second only to the dependency ratio in Japan. Such profound demographic changes will require a substantial reform of the pension system, which will have to be completed before the financial effects of the changes become disruptive.

The pivotal point in the discussion about pension reform is the question of how the prospective increase in the financial burden should be allocated between the contributors to the pension plans, the recipients of pension benefits, and the Federal Government. According to model calculations by the authorities, under present policies, contributions by employers and employees together would have to increase from 18.7 percent of gross wage and salary income at present to about 22 percent by the year 2010 even if the share of the Federal Government in the financing of pension payments were increased from the present ratio of 18 percent to 20 percent. If, on the other hand, the rate of contribution by employers and employees were limited to about 20 percent, financing by the Federal Government would have to rise to about 25 percent. These results suggest that it may be difficult to complete the reform of the pension system without a contribution from the recipients of pension benefits. In this respect, two areas of possible cost savings exist. First, pension payments could be linked to the development of net rather than gross wage and salary increases.[127] Second, the retirement age could be raised.

Medical Insurance

While the Government succeeded in keeping the rate of growth of aggregate social transfer payments below that of GNP, the cost of the medical insurance system continued to grow at a much higher rate. In 1981–86, expenditures on medical insurance increased at an average annual rate of 5 percent, compared with annual increases of 4 percent for total social transfer payments, 4½ percent for nominal GNP, and only 3½ percent for gross wage and salary incomes (Table A38). Thus, the numerous past attempts at containing medical costs were not fully successful.[128] After a lowering of the rate of contribution from 12 percent of wage and salary income in 1982 to 11.4 percent in 1984, the medical insurance fund ran deficits in 1984–86; consequently the contribution rate had to be increased again to an historically high level of 12.6 percent in 1987 (Table A39).

In November 1987, the Federal Government announced a plan for the reform of the medical insurance system to become effective in 1989; it consists of (1) a reduction in existing benefits, (2) the introduction of new benefits, and (3) measures to increase the cost efficiency of the system.

Reductions in benefits comprise the removal of death benefits for surviving family members, with transitory provisions for hardship cases; the coverage of only predetermined costs for medication and therapy (including spectacles and hearing aids) by health plans, so that overcharging by some providers of medical services will be excluded; the refunding of only 50 percent of the cost of dentures, but variations in coverage by plus or minus 10 percentage points are possible on a case-by-case basis; a limit (DM 15 a day a person) on the amount health plans can contribute to outpatient health cures; and the discontinuance by public health plans of reimbursing patients for the cost of taxi rides to their doctors or hospitals.

New benefits provided by public health plans comprise the coverage of costs for the home care of gravely ill people and preventive health care.

Measures to increase cost efficiency comprise a larger refund to those who need dentures and can prove that they have had regular prophylactic treatment; the refund of part of the contribution to those who did not file a claim during a particular year; an increase in the financial auditing of hospitals; and the termination of legislative differences in contributions and benefits for blue-collar workers, on the one hand, and white-collar workers, on the other. The most important effect of this last measure will be that blue-collar workers will also be able to leave public health plans when their income exceeds a certain level.

Altogether, the plan foresees only a minor increase in cost sharing by the consumers of health services, but it narrows the coverage of medical expenses to more basic services and improves incentives for preventive care. The plan will have a limited effect on the hospital sector, which is the fastest growing area of health expenditures; hospitals are administered by the federal states and municipalities and therefore

---

[127] At present, the formal link is still to gross income increases, but over recent years a discretionary element has been introduced with the result that pensions have in the event increased more in line with net incomes.

[128] Between 1976 and 1987, six laws were passed with a view to containing medical costs, see Seffen (1987, pp. D1–D10).

IV • STRUCTURAL POLICIES

Table 9. Financial Implications of the Health System Reform

(In billions of deutsche mark)

	1989	1990	1991	1992
Additional expenses due to new benefits	1.2	1.8	5.6	7.9
Savings due to reduction in benefits and higher cost efficiency	7.6	11.2	12.1	14.1
Net savings	6.4	9.4	6.5	6.2

Source: Ministry of Labor.

largely beyond the reach of the Federal Government.

The projected financial implications of the reform of the health insurance system are shown in Table 9. The estimated net savings could, if they materialized, eventually allow a reduction in the rate of contribution to the medical insurance system by ½–1 percentage point.[129] In 1988, however, the announcement of the plan has caused a surge in the demand for spectacles and dentures for which the insured will have to pay more after the reform has become effective.

Unemployment Insurance

Over the years, the unemployment insurance system has changed from an income support scheme for those who are temporarily out of work to a major instrument of labor market policy. In fact, while total expenditures of the unemployment insurance system have increased by about one fourth since 1983, expenditures on the "active labor market policy" have doubled. In particular, measures to educate or retrain the unemployed have been strengthened (in 1987, about 600,000 persons were enrolled in the education programs of the unemployment insurance system) and the employment creation program (*Arbeitbeschaffungsmassnahmen*) has been extended (in 1987, about 115,000 persons were sponsored through this program).[130] Moreover, compensation for short-time work (*Kurzarbeitergeld*) was generally extended to 24 months, and, for the steel industry, as of January 1, 1988, compensation for short-time work may also be paid to employees of companies that undergo restructuring programs. The maximum duration of these and other short-time work payments to the steel industry was extended to 36 months until the end of 1989.

All these programs have the objective of promoting economic adjustment and easing the associated costs for workers and employees. But they also contribute to a more rapid rise in expenditures than in the revenues of the unemployment insurance system, raising the prospect of further increases in the rate of contribution. For employers, however, larger contributions mean, in effect, higher labor costs, and are likely to price more marginal workers out of the market.

Indeed, as a result of the rapidly rising costs for unemployment insurance, but also for the public health and pension plans, and due to social legislation, the wedge between wage costs and take-home pay before direct taxes has risen substantially (Chart 15). The resulting high wage costs are one of the reasons for the high unemployment rate and underline the need for a comprehensive reform of the social security system with a view to containing social expenditures. They also raise the question whether alternative and less distortionary means of financing social expenditures should be used.

Trade and Industrial Policies

After several GATT rounds of trade liberalization, external tariffs for German manufacturing are relatively low; however, to conclude from this that external protection is correspondingly low would be wrong. As shown in Tables A40–A42, additional protection is afforded by nontariff barriers and tariff escalation. Indeed, a recent study by the Kiel Institute[131] argues that tariffs have increasingly been replaced by nontariff barriers, such as quotas, "voluntary" export restraints, and countervailing duties; total protection has increased since the Tokyo Round; and measures have become more selective, increasing protection in particular for a few sectors and industries (agriculture, coal mining, iron and steel, textiles, and clothing).[132]

In addition, subsidies have become an important element in the system of external protection in Germany. The financial assistance provided by the general government, measured on a national accounts basis, has by and large increased in line with nominal GNP between 1975 and 1987. At 2 percent of GNP, subsidy payments in Germany are close to the average of industrial countries; subsidies are larger in some European countries, notably France and Italy, but they are much lower in the United States and Japan (Ta-

---

[129] This reduction may, however, only be temporary as the aging population could raise the demand for medical services in the more distant future and, as a consequence, require an increase in the rate of contribution.

[130] Under this program, part or all of the wage of a worker who was previously unemployed is paid, for a specified period, by the unemployment insurance system.

[131] Donges and others (1988).

[132] In Table A40, the dispersion of protection across industries has been calculated. The coefficients of variation increase from less than ½ for tariff protection to 1 for total nominal protection and 2 for total effective protection. This gives an indication of the relative importance of nontariff barriers, including subsidies, and tariff escalation in the external protection of certain industries.

ble A43). The definition and measurement of subsidies varies, however, across countries and gives rise to disagreement even within countries. In the national accounts, for example, subsidies are defined as current income transfers. Thus, interest subsidies would be included, but not contributions to investment outlays, which are considered to be capital transfers. The biannual Subsidy Report of the Federal Government therefore uses a broader definition of subsidies than that of the national accounts, taking into account tax preferences and capital transfers.[133] Under this definition, although being substantially higher than under the national accounts definition, subsidies of the territorial authorities have actually declined relative to GNP between 1975 and 1987 (Table A44). But, since 1982, when the ruling coalition came into office, subsidies of the Federal Government have remained roughly unchanged at 1½ percent of GNP.[134]

About 40 percent of subsidies go to private households and the rest to the business sector.[135] The most heavily subsidized sectors and industries have been agriculture, mining, aircraft, shipbuilding, construction, and transportation (Tables A45 and A46). In the first three sectors, subsidies have increased strongly between 1985 and 1988, reflecting a policy that sought to protect these sectors from the appreciation of the deutsche mark during the period.

Agriculture

In 1986, agricultural value added accounted for almost 2 percent of GNP. The agricultural sector employed about 5¼ percent of the total employed civilian work force and agricultural products accounted for about 13½ percent of imports and 3½ percent of exports. Thus, agriculture plays a less important role in the German national economy than in France or Italy, but it is more important, at least as an employer, than in the United Kingdom (Table A47).

From the mid-1970s to the mid-1980s, the rate of growth of production and gross value added in agriculture was lower in Germany than in the EC on average (Table A48).[136] Despite the support given under the Common Agricultural Policy of the European Communities (CAP) and national aid to farmers, real agricultural incomes, measured as net agricultural value added at factor cost per manpower unit, have declined in Germany between 1975 and 1985 (Table A50). Most of this decline occurred in 1975–81, a time when real agricultural incomes rose rapidly in other EC countries, reflecting a decline in the productivity of German farmers; in 1982–85, the decline was much smaller and paralleled a similar decline in the EC as a whole.

Pricing policies under the CAP have kept producer prices in Germany at times substantially above world market prices (Table A51).[137] This has contributed to a significant rise in self-sufficiency for a number of important products (Table A52) and a rapid increase in the costs of the CAP (Tables A53 and A54). National subsidies to agriculture have also increased over recent years. According to the Federal Government's Subsidy Report, they rose from DM 2¾ billion or 7¼ percent of value added in 1982 to more than DM 5 billion or 17 percent of value added in 1987. In that year, national and supranational subsidies to agriculture amounted to almost half of the total value added in agriculture (Table A45).

In early 1987, the EC Commission proposed a price freeze for most agricultural products and somewhat lower prices for a selected subset for the marketing year ended in March 1988.[138] In addition, the Commission suggested a limit on market intervention and a change in the system of "green exchange rates," in which prices in ECUs are linked to the strongest currency in the European Monetary System. The Commission advocated a return to the previous system that was suspended in 1984 with a view to abolishing the green exchange rate system completely by 1992.[139] To raise additional revenue for the financing of the support mechanism for vegetable oils and fats, it was also proposed that a consumer tax on all imported and domestically produced vegetable and marine oils and fats be introduced. These proposals came against the background of mounting budgetary problems in the EC that were triggered by the sharp increase in spending on agriculture in recent years.

On July 1, 1987, EC ministers of agriculture reached agreement on the intervention prices for the marketing

---

[133] Some private researchers use even broader definitions and consequently arrive at higher estimates for subsidies in Germany.

[134] While subsidy payments have also remained largely unchanged at 5.4 percent of federal government expenditures since 1982, tax preferences have increased from 6.5 percent of total tax revenues in 1982 to 7.4 percent in 1987 (they are projected to increase further to 7.7 percent in 1988 but to fall from 1990 onward).

[135] The share of subsidies to households has declined from 44 percent in 1985 to an estimated 38½ percent in 1988.

[136] In 1960–80, the reallocation of labor from agriculture to other sectors, in particular services, proceeded at a similar pace in Germany to that in the other EC countries, but, since 1980, there has been a slowdown in Germany in reallocating labor. The share of agricultural employment in total employment declined by only 5 percent between 1980 and 1986 in Germany while it fell by about 12½ percent in the EC on average (Table A49).

[137] For a description of the institutional setting of the CAP and a discussion of its economic effects, see Rosenblatt and others (1988).

[138] A further freeze of agricultural prices in ECU terms was decided in mid-1988 for the marketing year 1988/89.

[139] See Rosenblatt and others (1988) for detailed information on the evolution and functioning of the agrimonetary system of the CAP.

IV • STRUCTURAL POLICIES

year 1987/88 (Table A55) and a change in the agri-monetary system (but not on the proposed introduction of a tax on oils and fats). As a result of this agreement real output prices continued to decline in 1987 both in Germany and in the EC on average (Table A56). Owing to an even steeper fall in input prices, however, the price/cost ratio improved. The decisions for the marketing year 1987/88 also included limitations on intervention for some products and a change in the agri-monetary system with a view to eliminating Monetary Compensatory Amounts (MCAs) and green exchange rates that differ from market rates by 1992.

Under an agreement reached in 1984, the German authorities were entitled to compensate farmers for price reductions related to the dismantling of existing positive MCAs. The compensation was set at 5 percent of sales for the period from July 1, 1984 to December 31, 1988, and to 3 percent of sales from January 1, 1989 to December 31, 1991, the day when all MCAs had to be abolished. These subsidies were expected to cost the Federal and the Länder Governments DM 18.4 billion altogether.[140] In order to divorce subsidies from production incentives, EC ministers of agriculture decided at their meeting in July to grant part of the compensation to German farmers in the form of income transfers independent of output, equivalent to 2 percent of estimated agricultural sales or about DM 1.2 billion in 1988 (with total compensation in 1988 estimated at DM 3 billion). The remaining positive MCAs are to be eliminated at the beginning of the marketing year 1989/90 and it was decided to fully compensate for this by means of national income transfers that do not affect production or competition in the Communities.

Despite these measures and an increase in the contributions by the member states from the equivalent of 1 percent to 1.4 percent of a harmonized value-added tax base, which came into effect in 1986, the budgetary situation of the Communities deteriorated in 1987. Indeed, a budgetary deficit of about ECU 3 billion was only avoided by national governments agreeing to finance a part of the expenditures incurred on account of agricultural price support through 1987. At a special meeting in February 1988, the European Council of Heads of State and Government reached agreement on a reform of EC finances and the CAP.[141] With regard to the CAP, the Council decided that expenditures for agricultural price support were to increase by not more than 74 percent of the increase in Community GNP, with outlays in 1988 budgeted at ECU 27.5 billion. Over the medium term this implies a considerable slowdown of the rate of growth of expenditures from earlier years. In order to achieve this target, existing agricultural stabilizers were reinforced and new measures were introduced. In particular, the Council set "guarantee thresholds" for cereals, oilseeds, and protein products and, in addition, imposed supplementary "co-responsibility" levies for cereals.[142] These measures supplemented arrangements for milk, sheepmeat, wine, sugar, tobacco, cotton, fruit, and vegetables, which had been adopted at a Council meeting on December 11–12, 1987.

At the request of the German authorities, a "land set-aside" scheme was introduced with the objective of limiting agricultural output. While the implementation of the scheme was made compulsory for member states, the choice of whether to participate was left to farmers. In order to qualify, a producer must set aside at least 20 percent of his arable land for at least five years. Farmers who set aside at least 30 percent will also be exempted from the co-responsibility levy on sales of cereals of up to 20 tons. Payments under the scheme will vary from ECU 100 to ECU 600 per hectare; they will be about 50 percent less if the arable land is used for fallow grazing or converted to certain types of protein plant production. The Community budget will contribute between 15 percent and 50 percent to these payments, at an estimated annual cost of ECU 200 million, with the rest coming from national budgets. It was also decided to introduce optional arrangements for promoting the early retirement of farmers. According to preliminary estimates by the Federal Government, the set-aside scheme will cost Germany about DM 417 million in 1989 and about DM 440 million annually by 1992; costs for the early retirement scheme are expected to increase from DM 141 million in 1989 to DM 615 million in 1992.[143]

Over recent years, the need to limit surplus production and contain the costs of agricultural support in the EC has become more urgent. However, although a number of changes have been made in the Common Agricultural Policy of the EC, agricultural price support has remained the principal instrument to achieve the objectives of the CAP, particularly that of providing stable and adequate incomes to farmers. But it is doubtful whether the present problems of agriculture in the EC can be resolved without a move away from this instrument toward more market-oriented solutions.[144]

---

[140] Since the compensation was formally treated as a value-added tax rebate, both the Federal and Länder Governments, which share value-added tax revenues, had to contribute.

[141] See Rosenblatt and others (1988).

[142] Co-responsibility levies are meant to let producers share in, or bear, the cost of price support. Guarantee thresholds serve the same purpose by penalizing surplus production through reduction in the intervention price in the period following the one in which the threshold has been exceeded.

[143] In Germany, a large number of farmers are close to retirement age.

[144] See Rosenblatt and others (1988).

Coal Mining

Germany, like most of the other members of the European Coal and Steel Community (ECSC), has a comparative disadvantage in coal mining: mining shafts have to reach a depth of up to 1,000 meters, labor costs for the industry are relatively high, and the economic costs of pollution are substantial. Nevertheless, over decades, the Government has kept this industry going by a combination of import restrictions and subsidies. The reasons for this policy range from security—the desire to secure the supply of energy—to regional policy—the industry is largely concentrated in one area of the Federal Republic.

Coal is still the most important domestic source of energy. More than 90 percent of the supply of coal is consumed by the electricity and steel industries. In 1987, 77½ million tons of coal were produced, down from about 80 million tons in 1985–86; however, because of reduced production in the steel industry, only 74 million tons were consumed so that the already high stocks increased further.[145] In 1987, the coal mining industry employed about 158,000 workers, down from almost 240,000 workers in 1975. In December 1987, representatives of the employees and employers in the coal mining industry as well as of the concerned Länder governments and the Federal Government agreed that annual production had to be reduced by between 13 and 15 million tons by 1995 with the implication that about 30,000 workers had to be laid off.

To soften the consequences of this adjustment, the authorities promised further financial support. Moreover, they agreed to increase subsidized exports of coal by 11 million tons to compensate, at least in part, for the decline of domestic demand. These decisions were taken against the backdrop of sharp increases in financial support to the coal mining industry over recent years. In 1980–87, the Federal Government has provided about DM 37 billion in subsidies to coal mining (including DM 21 billion through the *Kohlepfennig*, see below) and additional support has been given by Länder governments. Subsidization has been mainly through marketing programs, in which the principal users of German coal (producers of steel and electricity) are compensated for the difference between the domestic price of coal and the world market prices of coal and oil, respectively, by the Federal Government and the electricity consumers. The marketing program for the electricity industry is contractually guaranteed until 1995 with sales planned to increase from 40 million tons in 1987 to about 46 million tons by 1995. Similarly, the practice of compensating the steel industry for the use of higher-priced German coal has been guaranteed within certain limits until the year 2000.

In 1987, the difference between the domestic and world market prices for coal amounted to DM 155.35 a ton. The Government agreed to subsidize the use of coal in steel production by DM 148.85 a ton at a total cost of DM 3.1 billion (of which DM 2.4 billion was borne by the Federal Government and the rest by the State Government of North-Rhine Westphalia). The coal industry's own contribution was set at DM 5 a ton while the steel industry bore DM 1.5 a ton. For 1988, marketing support by the Federal Government is budgeted at DM 2.4 billion.

Subsidies to the electricity industry to compensate for the higher cost of generating electricity with German coal are determined by the difference between the price of German coal and the world market price of oil. Reflecting continuing weak oil prices and the appreciation of the deutsche mark vis-à-vis the U.S. dollar, the required compensation rose in 1987 to DM 5.6 billion, up from about DM 3 billion in 1986 and DM 2 billion in 1985. In order to finance this subsidy, a levy on the price of electricity is collected (the *Kohlepfennig*). As a result of the sharp increase in subsidies for the use of coal in electricity generation, this levy had to be raised in June 1987 from 4½ to 7½ percent. A much sharper increase was only avoided by authorizing additional government-guaranteed borrowing of up to DM 2 billion by the "coal equalization fund," whose purpose is to finance temporary imbalances between receipts from the *Kohlepfennig* and subsidy payments—the previous borrowing limit had been set at DM 0.5 billion. By the end of 1987, the liabilities of the fund amounted to DM 2.2 billion.

As in the case of agriculture, the marketing arrangements for coal imply, in effect, open-ended support—subsidy payments by the Federal Government or the electricity consumer depend largely on the world market price for oil and coal and the U.S. dollar exchange rate, and are therefore not subject to budgetary ceilings. At a meeting with representatives of the coal industry in December 1987, the Federal Government announced that it intended to rein back public support for the use of coal in electricity production to the level that existed before the decline in oil prices and the depreciation of the U.S. dollar. The Government's plan is to limit compensation to the electricity industry so as to allow a gradual reduction of the *Kohlepfennig*, which at present stands at 7¼ percent. Despite the decline in support, the Govern-

---

[145] Imports of coal are subject to a quota and discouraged by the high subsidies for the use of German coal. In 1987, 7.4 million tons of coal were imported, less than half of the amount allowed under the quota. Most of these imports were offset by subsidized exports of German coal to other members of the ECSC.

ment expects the electricity industry to purchase the contractually guaranteed amounts of coal. So far, the latter has resisted the Government's plan as it would shift the exchange rate and oil price risk from the Government to the electricity industry.

Steel

The responsibility for industrial policy in the steel sector lies with the Commission of the European Communities. According to the treaty of the ECSC, the Commission, with the consent of the Council, has the power to fix output prices, impose production quotas, and regulate trade with third countries, if this is deemed necessary to avoid a crisis situation. National measures to support the steel industry, in particular subsidies, are not allowed under the treaty. Since 1975, the European steel industry has had considerable excess capacity (Table A57).

As a result, employment in the steel industry of the Community of Nine declined by more than 50 percent between 1973 and July 1987 (Table A58). National governments, in clear violation of the ECSC treaty, initially reacted to the political pressures from the steel industry and trade unions by granting subsidies. To contain an emerging competition between governments seeking to support their own steel industries, a system of production quotas was introduced in 1980. Shortly thereafter, a "subsidy codex" was established with a view to surveying and limiting the burgeoning national subsidies.[146] Under this policy, which was initially scheduled to expire at the end of 1985, steel production capacity was reduced by about 30 million tons, or 25 percent, but it was estimated that about 20–25 million tons of excess capacity remained.

In 1986, direct subsidies had to be ended[147] but production quotas remained in force for several products through 1987 and trade in steel between the EC and other countries continued to be regulated by a number of bilateral agreements.[148] The Commission proposed in September 1987 that, if the quota system were to be prolonged beyond the end of that year, it would have to be accompanied by firm commitments to reductions in excess capacity by the companies and governments concerned. In the event, these commitments were not made, but the Council of Economic Ministers decided to extend the quota system, though in a somewhat modified form (Table A60). For coils, cold-rolled sheets, quarto plate, and heavy sections, quotas were to continue for the first half of 1988. Moreover, for hot-rolled coil and cold-rolled sheet, quotas were to remain in force through the end of 1990 provided that commitments were made by June 1988 to reduce capacity by 7½ million tons—about three fourths of the estimated surplus capacity in these products. Similarly, for quarto plate and heavy sections, the extension of quotas through 1990 was made subject to commitments to a reduction of the estimated excess capacity by 75 percent. As proposed by the Commission, quotas on wire rod and merchant bar were ended with effect from January 1, 1988. Thus, production of a little less than half of products covered under the ECSC treaty was liberalized.

Steel output in the first four months of 1988 was significantly above the level of a year earlier and exports were stronger. This weakened the readiness of firms and governments to actively pursue capacity reductions within the given time limit. Therefore the Commission proposed not to prolong the quota system beyond June 30, 1988. Despite the end of the quotas, the Commission maintains the statistical monitoring system for all products except coated sheet and reinforcing bars and continues to publish quarterly forecasts of production and deliveries. Concerning the problem of subsidization, it has been argued that some countries have continued to extend hidden operating subsidies in the form of subsidized credits to steel mills by publicly owned banks. The Commission took action to examine whether these measures are legal. The concerns about hidden subsidies are shared by the German authorities.

In Germany, the problems of the steel industry have reinforced the depressing effects of the decline of coal mining on one of the country's most important industrial areas, the Ruhrgebiet. At a special meeting on the economic problems of this area in February 1988 (*Montankonferenz*), the authorities promised a package of regional financial support of DM 1 billion, financed by the Federal Government and the state government of North-Rhine Westphalia. The Federal Government also announced additional investments by public enterprises to improve the infrastructure of the region and incentives to attract private investment. Trade unions, however, were little impressed by the decline of the steel industry and the regional problems of the Ruhrgebiet. The wage settlement for the iron and steel industry, reached only two days after the above-mentioned special meeting, foresaw a 2 percent increase in wages with effect from March 1, 1988, and

---

[146] Nevertheless, subsidies continued to flow. In 1980–85, about DM 87 billion was transferred from national governments to the EC steel industry (Table A59). At about DM 56 a ton, the steel subsidy in Germany amounted to about one fourth of that in other major producers. The highest subsidies were paid in Italy, followed by Belgium, France, and the United Kingdom.

[147] Financial aid to companies that closed down plants, however, continued to be permitted.

[148] Bilateral agreements have governed EC trade in steel with third countries since 1978. For a description of recent changes in these agreements see Kenward (1987).

an additional 2 percent with effect from August 1, 1989; the contract period will last until October 31, 1990. In addition, with effect from November 1, 1988, the weekly working time will be reduced by 1½ hours to 36½ hours; as no corresponding reduction in wages will take place, earnings per man-hour will rise by 4.1 percent. As a result, labor costs in the steel industry are expected to increase by about 2½ percent in 1988 and by more than 4½ percent in 1989.

Shipbuilding

Since 1976 there has been global excess capacity in shipbuilding. Despite the recovery of the volume of international trade from its trough during the last recession, the demand for freight capacity has remained weak. Hence, the demand for new ships has remained low and the competition for available orders has increased. A substantial reduction in world shipping and shipbuilding capacity is called for. Given the increased low-cost competition from the newly industrialized countries, the share of EC countries in the reduction of worldwide capacity will probably have to be more than proportional.

Despite the fall in capacity and employment that has taken place (Tables A61 and A62), the EC Commission has estimated that capacity will have to be cut by one third from its 1985 level, with a greater-than-commensurate decline in employment because of expected increases in productivity. This reduction would enable the remaining installations to operate at 70 percent of available capacity through 1990, with the possibility of returning to an 80 percent rate of capacity utilization thereafter. The Commission intends to foster an orderly adjustment in this industry by limiting national production subsidies.

In 1980–87, the Federal Government granted the shipbuilding industry about DM 1½ billion in subsidies. For 1988, the Government expects to provide an additional DM 280 million. According to the current EC directive on aid to shipbuilding, which was adopted in early 1987, government aid is limited to 28 percent of the contract value.[149] This ceiling is to be reviewed annually and is expected to fall as adjustment proceeds. Investment incentives may be granted only in support of restructuring programs that eventually reduce capacity. Aid to defray the costs of closure, such as payments to redundant workers or yard redevelopment is allowed, but must be consistent with the restructuring.

Other Industries

Other industries subject to pervasive government intervention are textiles, clothing, and commercial aircraft. Trade in textiles and clothing is subject to the Multifiber Arrangement (MFA IV) that was renewed for a third time in 1986 and is to remain in force until 1991. It has been argued that protection of the textile industry has reduced consumers' welfare in the importing countries and depressed export earnings of developing countries.[150] In the last part of this section, evidence is presented that protection of the textiles and clothing industry contributes to the present adjustment problems of the German economy. Nevertheless, the German authorities supported the renewal of the MFA which, they believe, will facilitate a further orderly adjustment of the German textile industry.

The development and, until recently, the production of commercial aircraft has been supported by subsidies from the Federal Government. Through 1986, Federal Government support for the commercial aircraft industry (mainly Airbus) amounted to DM 4.9 billion, of which DM 0.4 billion was granted in the form of marketing support and DM 0.7 billion as direct production subsidies. The remainder was given as loans for the development of the Airbus models A300, A310, and A320 (a significant part of which was subsequently forgiven). These subsidies are channeled through Deutsche Airbus GmbH, which owns 37.9 percent of the European consortium Airbus Industrie.[151] In addition, the Government guaranteed DM 3.1 billion of commercial credits to the aircraft manufacturer. In 1987, the Federal Government was forced to take over part of this debt, in the amount of DM 1.9 billion to be amortized in the period 1988–94, as Airbus Industrie experienced financial difficulties triggered by the fierce international competition and the decline of the U.S. dollar. Furthermore, the federal authorities agreed to provide an additional DM 3 billion over the 1988–96 period for the development of new aircraft models (A330 and A340). The German aircraft industry will repay this assistance depending on the development of sales.[152]

Government support for the European aircraft in-

---

[149] See Council Directive of January 26, 1987, on aid to shipbuilding in *Official Journal of the European Communities* (Paris), No. L 69 (March 12, 1987), pp. 55–60.

[150] See, for example, Keesing and Wolf (1980); Spinanger and Zietz (1986); and Kirmani, Molajoni, and Mayer (1984).

[151] "Deutsche Airbus" is a subsidiary of the largely privately owned arms and aircraft manufacturer Messerschmidt-Bölkow-Blohm.

[152] Since airplanes are priced in U.S. dollars, the exchange rate of European currencies vis-à-vis the dollar is an important factor for the profitability of Airbus Industrie. For the models A300 and A310, cost calculations were based on an exchange rate of DM 2.30 per U.S. dollar; for the A320, the rate was DM 2 per U.S. dollar. A depreciation of the dollar below these rates implies the need for additional subsidies for aircraft production.

dustry has led to a conflict with the U.S. authorities who have argued that this practice is in violation of GATT Articles.[153] Also, the 1987/88 Annual Report of the German Council of Economic Advisors, while acknowledging that there may be good reasons to protect infant industries, warned against institutionalizing another public support scheme for an internationally uncompetitive industry. The German authorities have justified the subsidies to the aircraft industry by reference to the positive external effects of this industry on research and development in other areas, and on the grounds that the international market for commercial aircraft is monopolistic in structure. Moreover, they have argued that U.S. manufacturers have been indirectly supported by military contracts. The authorities have also pointed to the relatively low share of Airbus Industrie in the world market (20 percent), which should mitigate the U.S. concerns; they have expressed their confidence that Airbus will become economically viable once it is able to offer a full fleet of airplanes. The German authorities and their EC partners have therefore encouraged greater participation of financially strong private European companies in the industry and collaboration with another manufacturer in the development and production of certain aircraft models.

Some of the trade and industrial policies described in this section have led to serious distortions in the economy. They have contributed to the freezing of resources in inefficient uses and thereby lowered growth and employment in Germany. This should be reason enough for change. But, by encouraging exports and discouraging imports, they have also helped to create Germany's large trade surplus, which has elevated the importance of these policies to the international level.

Privatization and Deregulation

When the present coalition government assumed office in 1982, the privatization of government holdings in industry and the deregulation of the economy were important points in its economic reform program. So far, however, progress in these areas has been mixed; privatization by the Federal Government has proceeded largely according to plan but much remains to be done on privatization of municipal services and deregulation.

The unsuccessful diversification policy of publicly owned enterprises in the 1970s and the recession in the early 1980s contributed to a severe deterioration of the financial situation of a number of publicly owned enterprises; in the event, losses had to be financed out of tax revenues. On October 26, 1983, the Government decided to develop a master plan for the phasing out of direct federal participation in industry; this plan was approved on March 26, 1985. According to the plan, government participation in industry was to be limited to cases where a clear public interest existed and where no other means were available to achieve the desired objectives. Thus, government activities were to complement rather than substitute for private activities. With regard to the privatization of government holdings and entrepreneurial activities, it was decided to proceed gradually in order to avoid disruptions. Also, loss-making government-owned enterprises had to become profitable again before privatization could be considered.

By 1985, losses of most government-owned enterprises had been cut significantly. By the end of 1987 a number of government holdings and activities had been privatized. In particular, the Federal Government reduced its participation in VEBA AG, the giant chemicals conglomerate, in January 1984 from 43.75 percent to 30 percent and sold the rest of its shares in March 1987. In June 1986, the Federal Government sold 40 percent of VIAG, a holding company for utilities, aluminum, and chemical companies, and in October 1986, it sold 45 percent of IVG, an industrial holding company. The Government intended to sell its shares in the Volkswagenwerk (VW) at the end of 1987; after the stock market crash, however, the sale was postponed until 1988. In early 1988, the Federal Government sold its remaining shares in VW and VIAG AG and was preparing the sale of its participations in Deutsche Pfandbriefanstalt and Deutsche Siedlungs und Landesrentenbank;[154] the federal railways intend to sell their holding in Deutsche Verkehrs-Kredit-Bank AG. It is expected that government revenue from privatization during 1984–88 will total almost DM 7 billion. In all, good progress has been made toward completing the privatization program of the Federal Government but ample room remains for the privatization of municipal services and deregulation.

In March 1988, the Government appointed an expert commission to investigate the economic costs of regulation and to propose steps toward deregulation of the economy. The report of this commission is expected in 1990. The need to reduce rules and regulations, which act to limit the flexibility and dynamism of the German economy, has been reinforced by the objective of fully liberalizing trade in services within the EC by 1992. Among the industries most affected

---

[153] See Kelly and others (1988).

[154] The Government will, however, retain its shares in Saarbergwerke and Salzgitter, two loss-making coal and steel companies.

by the completion of the internal market in Europe are the financial services, telecommunications, and transportation industries; other important industries that are subject to pervasive government regulations are retail trade and professional services. The prevailing regulatory system in these industries and planned changes are briefly reviewed below.

Financial Services

The banking industry is exempt from the provisions of the antitrust law (*Kartellgesetz*), which prohibits restraints on competition, price collusion, and other collusive agreements.[155] However, the supervisory authorities, the Federal Cartel Office (*Bundeskartellamt*) and the Federal Supervisory Office for Credit Institutions (*Bundesamt für das Kreditwesen*) have to approve any agreements that limit free competition among banks. The Federal Banking Act contains regulations with respect to market entry, the structure of balance sheets, and the exposure to credit risks. The Supervisory Office ensures that these regulations are observed.

The equity markets have historically played a smaller role in the raising of funds for the business sector than in some other major countries with the result that the business sector is highly leveraged by international standards.[156] Only some 2,200 of German enterprises are in the form of public companies; in contrast, some 5,300 companies in the United Kingdom were public companies in 1983. At the end of 1987, the shares of 574 domestic companies were quoted on the German stock exchanges (including the new "regulated" market discussed below) compared with 2,135 on the London stock exchange, 1,147 on the Toronto stock exchange, and 650 on the Paris stock exchange.[157] Table A63 provides some international comparisons of the capitalization of equity markets in the Group of Seven countries.

The small role of equity markets in financing the enterprise sector in Germany relative to some of the other Group of Seven countries reflects a number of factors. First, there has apparently been a lack of competition in underwriting activity as underwriting has been dominated by a small number of large banks.

Apart from the cost of issuing shares, there has been a long tradition of "house banks" acting as providers of long-term funds. Moreover, there are some additional factors that act as a disincentive to public companies: additional disclosure requirements, heavier auditing costs, and a requirement that employees participate on the supervisory board. Finally, private pension plans are less active in the stock market than in some other countries. This reflects, inter alia, the comprehensive nature of the social security system and the fact that company-run pension plans are allowed to retain contributions within the company (companies usually provide for future pension liabilities on their balance sheets).

In the 1980s, there has been a good deal more interest in the stock market as a source of funding than in the 1970s; in part this reflected the buoyant prices in the first half of the 1980s but it also appears that banks have become more aware of their exposure to highly leveraged enterprises and have been encouraging companies to boost the equity component of their financing.

There have also been significant legislative changes that have taken effect since the beginning of 1987. A new regulated market has been created (it commenced operations on May 4, 1987). The market provides the benefits of a stock exchange listing but with fewer reporting requirements and lower admission charges than the "official" stock market. Moreover, in contrast to the official market, institutions other than banks are permitted to act as underwriters on this new market;[158] this, it is hoped, will promote competition in underwriting activity. The new market attracted a significant share of new issues during 1987 and a large number of companies transferred their shares to the new market from the "regulated free" market and the "unregulated" market.[159]

A second legislative initiative introduced a new type of equity investment company, *Unternehmensbeteiligungsgesellschaft* (UBGG), to provide greater access to equity finance for small and medium-sized companies; UBGGs acquire participation in such enterprises and must themselves go to the official stock market or the new regulated market within ten years of establishment. Many leading banks and some insurance companies have set up UBGGs or announced their intention to do so. Despite the initial flurry of activity, it is too early to get a clear impression of how successful UBGGs and the new regulated market segment will be in getting equity funding to small and medium-sized

---

[155] The antitrust law and the exemptions granted to certain services industries are at present under review. The Government received the report of an expert commission in May 1988. On the basis of this report it will decide on the extent to which a reform of the law is warranted.

[156] Calculations by the OECD in 1986 showed that valuing assets at replacement cost, the debt/equity ratio in Germany was about twice the level as that in the United States and the United Kingdom, though about the same as that in Japan. See OECD (1986, p. 49).

[157] Data come from the Federation of the German Stock Exchanges (1988).

[158] The stock exchanges judge the qualifications of an institution to bring a firm to this new market segment.

[159] The regulated free market was abolished at the beginning of May 1988; thus, there are again three segments to the German equity market.

IV • STRUCTURAL POLICIES

enterprises or in providing an attractive range of new assets on the stock market. The eight stock exchanges in Germany have formed a working group to explore how the role of the stock market in Germany might be further enhanced.[160]

Despite the various institutional and regulatory changes over the past few years, there remain a number of factors impeding the further development of the capital markets in Germany. The most notable of these is the existence of a turnover tax on all secondary market dealings in bonds and equities.[161] This tax, while affecting the competitiveness of all financial markets in Germany, has particularly impeded the development of a market in short-term negotiable paper.

Moreover, the withholding tax on interest income as of 1989 will place an additional tax burden on German residents who were previously evading tax on their interest earnings and on foreign investors who cannot take advantage of exemptions under tax treaties.[162] In addition, a significant part of the yield of the withholding tax is expected to come from accumulated interest on life insurance policies, which was previously tax exempt.[163]

In light of the effective increase in the taxation of interest earnings (as well as the additional paperwork for investors), critics of the withholding tax argued that it would be quickly capitalized in the market thus either reducing the tax base (by drawing investors and borrowers to external financial markets) or raising the cost of public borrowing. It was generally expected that yields on bonds subject to withholding would rise by perhaps half the rate of withholding (30 basis points or so on average for public sector bonds) relative to foreign deutsche mark bonds, which will not be subject to withholding. In effect, however, it appears that the major part of the proposed withholding tax has been reflected in the relative yields of foreign and domestic

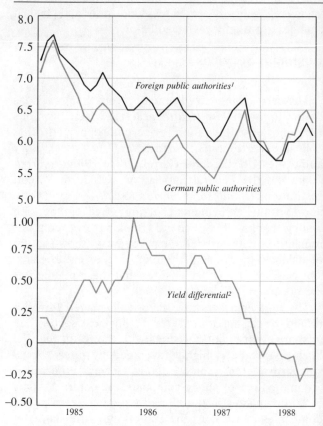

Chart 21. Federal Republic of Germany: Yields on Deutsche Mark Bonds Issued by Domestic and Foreign Public Authorities

(In percent a year)

Sources: Deutsche Bundesbank, *Monthly Report, Supplement 2*, various issues; and International Monetary Fund, *International Financial Statistics*, various issues.
[1] Public authorities from industrial countries.
[2] Yield on foreign bonds minus yield on German bonds.

deutsche mark bonds.[164] Chart 21 shows the development over the past few years in the yields of foreign deutsche mark bonds issued by foreign public authorities and the yield on bonds issued by German public authorities.[165] In the first nine months of 1987, the average yield on German public authority bonds was over 50 basis points lower than the average yield on issues by foreign public bodies. After the announcement of the withholding tax, this differential was almost

---

[160] For more details, see the Federation of German Stock Exchanges (1988).

[161] This tax (*Börsenumsatzsteuer*) is levied at a basic rate of 0.25 percent on secondary dealings in shares and bonds, with lower rates applying to certain transactions, including residents' transactions with foreigners. Transactions between banks and transactions involving bonds of the federal and state governments are not subject to the tax.

[162] In the case of foreign investors who can take advantage of tax treaties, this would involve having their application certified by their domestic tax authorities, effectively alerting them to the foreign source income. Thus, to the extent that such investors were previously evading taxes in their home country, there would also be an additional tax burden on them.

[163] The withholding tax on such policies would apply to interest earnings related to the excess of the actual interest rate over 3.5 percent; no refunds would be available for taxes withheld. The yield is expected to be in the region of DM 1¼ billion, or more than one quarter of the total expected revenue yield from the withholding tax.

[164] There is speculation that this reflects fears in the market that the rate of withholding might be increased at a later stage; indeed, the academic advisory committee to the Ministry of Finance had recommended a withholding rate of 25 percent.

[165] These are average yields for the bonds issued by the two groups of borrowers. The actual levels of the yields for the two groups may not be strictly comparable, for example, because of differences in the maturities of outstanding bonds or differences in bond quality; however, changes in the yield differential between these two groups should be able to give an indication of the impact of the withholding tax.

entirely removed and indeed in August 1988 the foreign issues were trading about ¼ of 1 percentage point lower than the German issues.[166]

Apart from these tax factors, a number of institutional factors remain that affect the competitiveness of the German financial markets. The principal financial instruments still not permitted are foreign-currency-denominated bonds, though residents are allowed to issue or acquire such bonds on external financial markets. While certain types of options are allowed in the German markets, such instruments are little developed, and futures contracts have not yet been admitted to trading; it is suggested by some that this situation results in more market volatility than otherwise might be the case since it restricts the hedging opportunities open to German investors.[167] In the stock market, the cost of bringing an issue to market is said to have been influenced by a lack of competition in underwriting activity, as transactions in the official market must be undertaken through a bank operating in Germany; it is hoped that the growing importance of foreign banks in the German capital markets will enhance competition in underwriting. The domestic issue of bonds by enterprises is in effect discouraged by approval procedures, security requirements, and tax considerations;[168] as a result bond issues of German enterprises are now almost all in the foreign bond segment of the deutsche mark bond market.[169]

The insurance market is divided into different business areas. Thus, an insurance company offering, for example, private health insurance is barred from offering car insurance. As a result, the supply of specific insurance services is highly concentrated and competition exists in only very few markets. In 1982, the share of the ten largest suppliers was about 77 percent in health insurance, 49 percent in life insurance, and 31 percent in the more competitive accident and damage insurance market.[170] As regulation prevents any individual insurer from offering a complete range of coverage, large conglomerates have been created which, because of regulations limiting price competition, exist along with numerous small insurers.[171]

As in the case of the banking industry the general prohibitions against cartels and price collusion do not apply to the insurance industry. The Federal Cartel Office is charged with ensuring that no abuse of market power occurs and the supervisory agency (*Bundesamt für Versicherungswesen*) enforces the insurance law (*Versicherungsaufsichtsgesetz*). The main regulations under this law concern market entry, product design, insurance premiums, distribution of profits, and the structure of assets of the insurance companies. The opening of an insurance company requires the approval of the supervisory agency and is subject to certain conditions, such as a minimum level of equity, a detailed business plan, and the assurance that no other forms of business are pursued by the applicant. The approval, when given, applies only to one type of insurance service. At present, insurance companies from other EC countries have to open a local office under German law if they want to offer their services in Germany; this rule has kept the market share of foreign competitors limited.

The terms of the insurance offered have to be approved by the supervisory agency and are standardized within each sector; the only exemption is private health insurance for which a number of different services can be bought. Premiums charged for about 70 percent of the overall insurance volume are subject to approval by the authorities. Even in the remaining cases, the cartel office checks the prices for their perceived adequacy and the respective insurance associations recommend prices to their members. In the

---

[166] Similar effects of the proposed withholding tax are implied by some recent issues of foreign deutsche mark bonds by prime international borrowers; in some cases yields on such issues have been up to ½ of 1 percentage point lower than yields on German Federal Government bonds of similar maturities. It does not follow from the above analysis that yields on domestically issued deutsche mark bonds have increased by ½ of 1 percent or so above what they would have been without the proposed withholding tax; the withholding tax may also have pushed down yields on foreign deutsche mark bonds in light of the higher demand for such bonds as a result of the withholding tax.

[167] This is one of the areas being examined by the working group formed by the German stock exchanges. Some legal impediments currently affect the operation of forward markets in Germany. First, only registered traders, stock market operators, and nonresidents can be legally bound to a forward transaction; under the Civil Code, forward transactions are regarded as gambling and a private resident can withdraw from a forward contract by making a gambling plea. Second, insurance companies and investment companies are not allowed to participate in forward market trading. Legislation is currently planned that would establish the legal basis for futures trading and would amend the Civil Code to make forward market contracts binding on all, and plans are under way to open a futures exchange toward the end of 1989 or in early 1990.

[168] Most domestic bond issues (certain bonds with equity characteristics are an exception) have to be approved by the Federal Government to ensure that the borrower is sufficiently sound. Moreover, the issuer of the bond usually is required to provide satisfactory collateral against the bond issue. This in general provides little difficulty for banks since many of their bond issues are backed by liabilities of the federal, state, or local governments or by mortgages; it is, however, a disincentive to nonbanks issuing bonds on the domestic segment of the deutsche mark bond market. Finally, domestic bond issues (but not foreign bond issues by external affiliates) are part of the base on which the trade tax (*Gewerbesteuer*) on capital is levied.

[169] These issues of foreign deutsche mark bonds are channeled through external affiliates or subsidiaries, which act as a "foreign" issuer; usually the issue is guaranteed by the domestic parent company. The size of domestic industrial issues is negligible.

[170] See Soltwedel and others (1986, p. 25).

[171] Of a total of about 800 insurance companies that existed in 1980, about 280 had market shares below 0.1 percent, while the largest conglomerate accounted for about 45 percent of the market (ibid., pp. 26–27).

case of life insurance, for instance, which accounts for 40 percent of the overall insurance market, these practices have led to uniform and excessively high (gross) premiums. At least 90 percent of the profits of insurance companies have to be redistributed to the customers and, in the absence of price competition, insurers sometimes compete on the basis of these reimbursements (this has resulted in the distribution of 98 percent of profits on average). After the completion of the internal market, it is envisaged that insurance companies from other EC member states will be allowed to offer their services in Germany without having to comply with German law. Thus, a significant change in the organization of the German insurance industry will have to take place.

Telecommunications

The Federal Post Office holds the legal monopoly for telecommunications services and thus regulates the market for related appliances. Critics of this institutional arrangement have argued that it has contributed to keeping the price for telecommunications services higher than in countries where private companies operate[172] and that it has in effect limited the access of foreign manufacturers of equipment to the German market.

In June 1987, the EC Commission issued a "Green Paper on the Development of the Common Market for Telecommunications Services and Equipment" that outlined the reforms required for a free common market in telecommunications services. For 1988, the Commission demands measures to liberalize the market for equipment and general guidelines for the access to telecommunication networks. Not much later, in September 1987, an expert commission appointed in 1985 by the Federal Government suggested a modest reform of the German telecommunications sector. It is planned to split the Federal Post Office into three public enterprises responsible for the mail service, the banking services provided by the post office, and the telecommunications services. The telecommunications company (TELEKOM) will retain the monopoly over the terrestrial transmission network and real-time voice telephone services. All other services (e.g., data transmission) will be open to private competition. Furthermore, the market for telecommunications equipment, including telephones, will be liberalized. Draft legislation on these reforms was submitted to Parliament at the end of June 1988, and the Reform Act is expected to become effective in 1989.

Transportation

Like financial services, the transportation industry is not subject to the antitrust provisions of the cartel law. Reflecting the diversity of this industry, regulations differ widely, but price and market entry restrictions apply to most transportation services. In the following paragraph, the most important rules and regulations for road transportation are reviewed briefly.

Long-distance trucking (more than 50 kilometers) is subject to licensing. Licenses are valid for eight years and can be renewed. Market entry is only possible if an existing license can be acquired or if the Ministry of Transportation increases the overall number of licenses. New licenses are assigned by the authorities but there is a secondary market for existing licenses. It appears that this system has contributed to monopoly rents accruing to the licensed companies. In 1984, for instance, for 2,100 new regional licenses more than 9,100 applications were received. Moreover, prices paid for licenses in the secondary market (up to DM 250,000) confirm that there are considerable rents in trucking.[173] Prices for long-distance transportation are set and enforced by the authorities (*Tarifkommission für den Strassengüterfernverkehr* and *Bundesanstalt für den Güterfernverkehr*). It has been estimated that the regulated prices in Germany are about 30 percent to 40 percent above free market prices, giving rise to some DM 12 billion annually in additional costs to German industry.[174] Cross-border trucking is also subject to licensing and, in part, to price regulation. "Community quotas" fix the number of trucks that are allowed to move freely in the EC and bilateral quotas regulate trucking between two member states. For the EC as a whole, about 54 percent of transportation services are subject to bilateral quotas and a further 16 percent to Community quotas; the remaining 30 percent are not subject to quotas.[175]

A full liberalization of the EC trucking industry is intended by 1993. The German authorities, however, have raised objections to the unconditional abolition of the quota system as they fear strong competition

---

[172] Indeed, Institut der deutschen Wirtschaft (1988) found that telecommunication costs in Germany were about 22 percent above the average of 11 industrial countries. They were about 171 percent above costs in Sweden, where telecommunications services were cheapest.

[173] Soltwedel and others (1986, p. 220). Long-distance trucking by manufacturing companies themselves (*Werksverkehr*) has increased considerably over recent years despite the requirement that company-owned trucks must not transport other loads and are therefore often forced to return empty from their destinations of delivery. This can be taken as an indication that prices charged by transportation companies are excessive.

[174] See Bangemann (1987).

[175] Only a third of this 30 percent, however, is accounted for by commercial trucking companies; the remainder consists of suppliers' own trucking services.

for their trucking industry, particularly from the low-cost suppliers in Belgium and the Netherlands.[176] They consider the harmonization of taxes and regulations for the protection of workers as a precondition for the liberalization of the trucking industry. Thus, when on June 20, 1988 the EC Council of Ministers decided on an increase in Community quotas for the trucking industry by 40 percent each year in 1988 and 1989, it also introduced some measures to harmonize social and technical regulations. A decision for a further increase in quotas in 1990-92 is required by March 1990 and further harmonization measures that will accompany full liberalization will be decided upon before June 1991.

Passenger transportation is also regulated by the authorities. Approval of new services, for instance, is only given when the authorities detect a need for the service and when it does not infringe upon already existing services. Licensed companies have priority over newcomers and the Federal Railways have the right to claim any long distance bus service for themselves. Regular bus fares have to be approved by the authorities. In the taxi business, licenses are required for each individual car and the number of taxis is regulated by the local authorities so as to avoid "excessive competition." Similarly, taxi fares within the jurisdiction of a municipality are regulated by the respective authorities; they can be set freely, however, for rides beyond the municipal boundaries.

Retail Trade

A law regulates shop opening hours (*Ladenschlussgesetz*). Shops have to remain closed on Sundays and public holidays, as well as before 7:00 a.m. and after 6:30 p.m. on Mondays through Fridays; on Saturdays they have to close at 2:00 p.m., except on the first Saturday of each month when opening hours are extended to 6:00 p.m. Price competition is restricted by the "Fair Trade Law" that regulates rebates and special sales. The latter are restricted nationwide to two seasons of the year and to a certain period preceding the permanent closure of a retail business. The justifications given for these restrictions range from the protection of employees in retailing to the protection of small shops, a purpose which the retail laws would not seem to have achieved over the last two decades.

The Federal Government intends to extend permissible shop opening hours to 9:00 p.m. on each Thursday and introduced draft legislation to this effect. This plan already has met fierce resistance from trade unions and shop owners' associations, although the law regulating shop opening hours would not be abolished. Moreover, as the ceiling on total weekly opening hours would not be increased, shops remaining open late on Thursdays would have to reduce their opening hours at some other time in the week.

It appears that the debate in Germany about shop opening hours is sometimes beside the point. For instance, the argument has been made that an extension of shop opening hours would be impracticable because it would raise the costs to retailers while revenues would remain the same, as consumers would not buy more goods. But this argument overlooks the fact that retailing is a service that commands its own price (reflected in the retail margin). Clearly, a better retail service—that is, at more convenient hours in less crowded stores—should command a higher price (i.e., margin) than the present poor service. This should allow retailers to hire the needed additional personnel—creating more part-time jobs—and, perhaps, even to raise profits. It is surprising that shop owners' associations would support government regulation that curtails their members' business opportunities and restricts the quality of service that can be provided.

Professional Services

Services rendered by physicians, dentists, accountants, notaries public, tax auditors, etc., are also subject to extensive regulation. Entry into some of these professions by qualified persons is restricted by ceilings on the number of professionals active in a certain area. The conduct of business is often regulated by professional associations, in which membership is compulsory. The two most common regulations concern the fees charged and advertising. Price regulations range from administrative price-fixing (pharmacists) and setting of minimum prices (e.g., attorneys) to the setting of price ranges (e.g., architects, engineers, physicians, and dentists). Advertising is prohibited in many professions and even the information given about the services rendered is regulated by the professional associations (e.g., the size of a name plate at a physician's or dentist's office, etc.). Consequently, competition is limited and the price and quality of these services are less attractive than they could be. The intended end to restrictions on the free movement of professionals within the EC by 1992 is likely to require some adjustment of the regulations.

The myriad rules and regulations in Germany are an important force militating against structural change. They also contribute to an environment in which private initiative and economic dynamism are stifled.

---

[176] Trucking companies in these countries are subject to fewer regulations and have to pay lower taxes than those in Germany.

IV • STRUCTURAL POLICIES

Deregulation is needed, but, as explained earlier, it is an arduous political process.

Economic Outlook Under Alternative Policy Scenarios

Among economists, there is broad agreement that structural reforms in Germany would greatly improve economic performance and contribute to external adjustment. Indeed, model simulations carried out in a study by the Kiel Institute[177] indicated that, as a result of trade liberalization and subsidy reduction, output and employment in the protected sectors would decline, but the decline would be more than offset by an increase in output and employment in unprotected sectors where Germany has a comparative advantage. As a result, GDP could increase by as much as 6 percent and employment by 9 percent, or two million jobs, thus virtually eliminating unemployment. Furthermore, in a recent study by Fund staff, it was estimated that the elimination of protection afforded under the Common Agricultural Policy of the EC could increase German GDP and employment by up to 3½ percent and 5½ percent, respectively.[178] In the following discussion of the economic outlook (until 1991) for the German economy three hypothetical alternative scenarios prepared by the authors are examined.

The analysis is conducted in the form of an illustrative quantitative exercise. In interpreting the simulations it is important to keep in mind that the results are contingent upon numerous simplifying model assumptions and assumed parameter values. They are therefore more controlled laboratory experiments than forecasts of actual developments of the German economy over the next few years. In the baseline scenario, it is assumed that present policies continue, except where changes have already been announced. Specifically, the tax reform package is assumed to go into effect as planned, monetary policy is expected to continue to be based on potential output growth together with some acceptable and sustainable rate of inflation, and the real effective exchange rate of the deutsche mark is held constant. In the first alternative scenario a more market-oriented pricing policy in agriculture and coal mining (i.e., lower administered prices in these sectors relative to other prices in 1989) is superimposed upon the baseline macroeconomic scenario. The second alternative scenario traces the effects of dismantling tariff and nontariff protection in the iron, steel, shipbuilding, textiles, and clothing industries under the macroeconomic assumptions of the baseline projection. The third alternative scenario combines the policy assumptions of the baseline and the two alternative scenarios. All three assume that regulations in goods and services markets and rigidities in the labor market are reduced so that the economy adjusts flexibly to the changes in administered prices and levels of protection.

The Methodology

Projections of economic developments under alternative structural policy measures require a model framework that is capable of both projecting macroeconomic aggregates and tracing the microeconomic effects of policy-induced relative price changes. No such model exists for Germany: the available dynamic macroeconomic models are incapable of dealing with structural change while the models that are capable of dealing fully with relative price effects at a disaggregated level are comparative-static models. Thus, the present exercise combines a dynamic macroeconomic model with a comparative-static, multisector, microeconomic model of the German economy. The macroeconomic model, which is described in more detail in Appendix II, is used for the baseline projections. This model projects the development of domestic macroeconomic variables under given assumptions about macroeconomic policies and the external environment during the projection period. Since behavioral relationships in individual economic sectors are not modeled, this framework cannot be used to analyze the sectoral policies described above. For this purpose, a comparative-static, four-sector computable general equilibrium (CGE) model is employed that allows the calculation of the deviations of economic variables from the baseline simulation in response to sectoral policy measures (Appendix III).

The dynamic macroeconomic model and the comparative-static CGE model are combined into one tool for the projection of economic developments in Germany through 1991 in the following way. Assume that the vector of dependent macroeconomic variables (Y) is a function of a vector of exogenous macroeconomic variables (A) and a vector of exogenous structural variables (B). Thus,

$$Y = Y(A,B). \tag{1}$$

Transformation of equation (1) yields

$$dY/Y = E1 \cdot dA/A + E2 \cdot dB/B, \tag{2}$$

where $E1$ and $E2$ are vectors of elasticities of A and B with respect to Y.[179]

---

[177] Donges and others (1988).
[178] Rosenblatt and others (1988).

[179] A common assumption in economic forecasting with macroeconomic models is that dB equals zero, that is, that there are no

In order to combine the two models, an assumption has to be made about the time period that is needed until a change in the structural environment is fully reflected in the endogenous vector Y. Assume that the length of this period is t–0, then the change in Y between time point 0 and time point t can be approximated by a function of changes in exogenous macroeconomic variables A between t and 0 and the change in B, which is assumed to have occurred in the base period 0.

$$(Y_t - Y_0)/Y_0 = E1 \cdot (A_t - A_0)/A_0 + E2 \cdot \Delta B_0/B_0. \quad (3)$$

Equation (3) was used to project Y_t under certain assumptions about $A_0, \ldots, A_t$ and the change in B, with $E1$ given by the macroeconomic model and $E2$ determined by the CGE model. The time period needed for the changes in B to affect Y was assumed to be two years.[180]

Simulations

Table A64 summarizes the results of the baseline and alternative projections. The former are based upon the announced changes in the stance of fiscal policy together with a constant real exchange rate in 1990–91 and partner-country demand in line with recent World Economic Outlook projections. The results are predicated on the following specific assumptions. Government consumption and investment expenditures grow at a rate of 4 percent a year in nominal terms (i.e., about 1 percentage point below the rate of growth of nominal GNP in the baseline scenario). The ratio of tax revenue to GNP falls by a little less than 1 percentage point in 1990 reflecting the planned tax cut. Foreign markets for German goods grow at a rate of 4½ percent a year, in line with the estimates of the recent World Economic Outlook exercise. Based on these assumptions, real GNP and domestic demand are projected to increase at a rate of 2.7 percent and 2.9 percent a year, respectively, and the current account is projected to decline to 3.6 percent of GNP in 1991, down from 4.0 percent in 1989. The rate of unemployment is projected to fall to 7.2 percent in 1991 from 7.7 percent in 1989 owing largely to a stagnant labor force and a reduction in working hours, and consumer prices are expected to increase by around 2¼ percent a year on average during 1990–91.

Scenario 1 assumes a 4 percent drop in the prices of food products and coal relative to other prices for 1988. With the GNP deflator projected to increase by about 2 percent in this year, this assumption implies a nominal price cut of about 2 percent for these products. In addition, it is assumed that fiscal and monetary policies continue to be directed toward the nominal targets of the authorities' medium-term strategy so that the rate of growth of nominal domestic demand remains unchanged from the baseline scenario. In this scenario, consumer prices increase by about ½ of 1 percentage point a year less than in the baseline scenario. Real consumption increases against the baseline scenario owing to higher real income growth and real balance effects. Real investment is stronger as the return to capital rises and real government consumption increases because the authorities are assumed to follow unchanged nominal expenditure targets. Consequently, the rate of growth of real domestic demand and GNP rises, and the current account surplus falls vis-à-vis the baseline projection. Reflecting stronger GNP growth, unemployment is also lower.[181]

Scenario 2 assumes the elimination of tariff and nontariff barriers in the iron, steel, shipbuilding, textiles, and clothing industries; the tariff equivalent of these barriers has been estimated at about 32 percent.[182] As a result of this trade liberalization, inflation falls relative to the baseline scenario, real domestic demand and GNP growth increase (see above) and the unemployment rate and the current account surplus decline. Since export demand and import supply elasticities of the CGE model differ, the changes in export and import volumes elicit a response of the terms of trade, which is taken into account in the computation of the change in the current account balance. With growth of both export volumes and import volumes significantly higher in scenario 2 than in the baseline scenario and scenario 1, there is a decline in the terms of trade vis-à-vis the baseline projection and scenario 1, which contributes to the larger current account adjustment.

Scenario 3 combines the previous two scenarios with the assumptions of the baseline projection. Reflecting the lower rate of inflation, real GNP and domestic demand in 1990–91 are projected to grow at annual rates of 3.4 percent and 3.8 percent, respectively, allowing the current account surplus to decline to 2.8 percent of GNP. Employment is projected to increase by 1.4 percent a year and the rate of unemployment to decline to 5.3 percent by 1991. Owing to somewhat lower nominal GNP growth (4.7 percent

---

changes in the structural environment. Thus, projections depend only on the assumed changes of exogenous macroeconomic variables (A).

[180] A period of about two years appeared to be appropriate for various reasons. For instance, the historical experience of the two oil price shocks in the 1970s seemed to suggest that up to two years were needed until all the effects of the price changes had worked their way through the economy. Moreover, the trade equations of the macroeconomic model also indicated a two-year lag until relative price changes were fully felt in the economy.

[181] Note that in this and the following scenarios it was assumed that fixed real wages above full employment levels gave rise to classical unemployment.

[182] Weiss (1985).

instead of 5.1 percent in the baseline scenario), government revenues increase at a slightly lower rate. With the increase in nominal government spending determined by the medium-term fiscal plan and therefore unchanged from the baseline, the general government deficit is projected to reach about 1.4 percent of GNP in 1991 in scenario 3, slightly higher than the 1.2 percent of GNP in the baseline scenario.

The sectoral developments that underlie the projections of the three policy scenarios are presented in Table A65. Since sectoral developments are only implicit in the baseline projection, the table shows deviations of selected sectoral variables from the values they would have attained if there had been no structural policy measures. These results indicate that price restraint in the basic goods sector (agriculture and coal mining) and trade liberalization in the iron, steel, shipbuilding, textiles, and clothing industries would lead to a reallocation of resources from these sectors and industries to manufacturing and services industries. As a consequence, the achievement of all four major macroeconomic objectives (growth, external equilibrium, price stability, and full employment) would be facilitated.

Appendix I
Statistical Tables

Table A1. Federal Republic of Germany: Aggregate Demand[1]

	1987		1983	1984	1985	1986	1987
	At current prices		At 1980 prices				
	(In billions of deutsche mark)	(In percent of GNP)	(Percentage change from previous period)[2]				
Private consumption	1,119.6	55.3	1.7	1.5	1.7	4.3	3.1
Public consumption	396.8	19.6	0.2	2.4	2.1	2.4	1.6
Gross fixed investment	388.3	19.2	3.2	0.8	0.1	3.1	1.7
Construction	219.7	10.9	1.7	1.6	−5.6	2.4	0.1
Residential	104.1	5.1	5.5	2.0	−10.0	−1.1	−0.9
Business	74.8	3.7	3.1	3.6	−1.3	4.9	1.5
Public	40.8	2.0	−9.2	−3.1	−0.8	7.1	0.2
Machinery and equipment	168.7	8.3	5.6	−0.5	9.4	4.1	4.0
Final domestic demand	1,904.7	94.1	1.7	1.6	1.4	3.6	2.5
Stockbuilding	8.7	0.4	0.6	0.5	−0.5	0.2	0.4
Total domestic demand	1,913.4	94.6	2.3	2.0	0.9	3.8	2.9
Exports of goods and services	636.6	31.5	−0.5	9.0	6.7	−0.2	0.8
Imports of goods and services	526.8	26.0	0.6	5.3	3.7	3.7	4.8
Foreign balance	109.8	5.4	−0.3	1.3	1.1	−1.1	−1.1
Gross national product	2,023.2	100.0	1.9	3.3	2.0	2.5	1.7

Sources: Statistisches Bundesamt, *Volkswirtschaftliche Gesamtrechnungen* and *Wirtschaft und Statistik*, various issues; and Deutsche Bundesbank, *Monthly Report, Supplement 4*, various issues.

[1] Totals may differ from sums of components because of rounding.
[2] Data for stockbuilding and the foreign balance are contributions to real GNP growth in percentage points.

Table A2. Federal Republic of Germany: Labor Market Developments[1]

(In millions)

	1980	1981	1982	1983	1984	1985	1986	1987
Employment	**26.3**	**26.1**	**25.7**	**25.3**	**25.4**	**25.5**	**25.8**	**26.0**
Manufacturing	9.0	8.8	8.5	8.1	8.1	8.1	8.3	8.3
Transport and communications	1.5	1.5	1.5	1.4	1.4	1.4	1.5	1.5
Agriculture, forestry, and fishing	1.4	1.4	1.4	1.4	1.4	1.4	1.3	1.3
Tradables sectors	11.9	11.7	11.3	11.0	10.9	10.9	11.1	11.0
Percent of labor force	43.7	42.5	41.1	39.7	39.3	39.3	39.5	39.2
Services (including finance, government, nonprofit)	8.3	8.4	8.5	8.6	8.7	8.9	9.1	9.3
Trade	3.5	3.5	3.4	3.3	3.3	3.3	3.3	3.3
Construction	2.1	2.0	1.9	1.9	1.9	1.8	1.8	1.7
Nontradables sectors	13.9	13.9	13.8	13.8	13.9	14.0	14.2	14.3
Percent of labor force	51.0	50.8	50.2	50.0	50.4	50.4	50.5	50.9
Other[2]	0.6	0.6	0.6	0.6	0.6	0.6	0.6	0.6
Unemployment	**0.9**	**1.3**	**1.8**	**2.3**	**2.3**	**2.3**	**2.2**	**2.2**
Percent of labor force[3]	3.3	4.6	6.7	8.1	8.1	8.2	7.9	7.9
Labor force	**27.2**	**27.4**	**27.5**	**27.6**	**27.6**	**27.8**	**28.0**	**28.2**

Sources: Statistisches Bundesamt, *Volkswirtschaftliche Gesamtrechnungen*, various issues; and Deutsche Bundesbank, *Monthly Report*, various issues.

[1] Figures may differ from sums of components because of rounding.
[2] Including energy, mining, and cross-border employment.
[3] Deutsche Bundesbank definition, based on microcensus data.

Table A3. Federal Republic of Germany: National Income

	1970	1980	1986	1987	1983	1984	1985	1986	1987
	(Share of national income)				(Percentage change from previous year)				
National income	100.0	100.0	100.0	100.0	5.1	5.6	4.7	6.5	3.8
Gross income from employment	68.0	73.5	68.8	68.8	2.0	3.6	3.9	5.1	3.8
Direct taxes	6.8	9.8	9.9	10.3	4.2	6.4	7.4	3.1	8.2
Social security contributions	13.2	21.3	21.6	21.6	4.0	5.3	4.8	5.4	3.7
Employer	7.0	13.4	13.4	13.4	4.5	5.4	4.4	5.3	3.6
Employee	6.2	7.9	8.2	8.2	3.2	5.3	5.6	5.5	4.0
Net income from employment	47.9	42.4	37.3	36.9	0.5	1.9	2.4	5.4	2.6
Gross income from entrepreneurial activity and property	32.0	26.5	31.2	31.2	13.8	10.6	6.5	9.7	4.0
Direct taxes	6.1	5.4	4.9	4.5	−2.6	6.5	12.2	2.9	−4.0
Net income from entrepreneurial activity and property	25.9	21.1	26.3	26.7	17.7	11.4	5.5	11.1	5.4
Private households and organizations	21.0	22.2	25.7	25.8	7.2	12.6	6.4	3.8	4.2
Government[1]	0.4	−1.0	−1.5	−1.9	...	...	...	...	...
Retained profits[2]	4.4	−0.1	2.2	2.9	12.2	8.7	4.6	32.7	45.0
Household disposable income[3]	81.1	83.9	81.2	81.2	2.8	4.7	3.7	4.7	3.8
Sources:									
Net wages and salaries	45.0	42.4	37.3	36.9	0.5	1.9	2.4	5.4	2.6
Distributed profits and property income	21.7	23.4	26.9	26.9	6.3	11.9	6.2	3.6	3.8
Net transfers	14.1	18.1	20.5	20.7	2.2	1.1	2.6	4.0	4.8
Less:									
Unrecorded payments[4]	2.4	2.5	2.3	2.2	1.9	5.6	2.9	1.7	−2.7
Interest costs	0.6	1.2	1.2	1.1	−7.2	−0.5	2.1	−1.2	−3.9

Sources: Statistisches Bundesamt, *Wirtschaft und Statistik* and *Volkswirtschaftliche Gesamtrechnungen*, various issues.
[1] Figures for government for 1983–87 are negative.
[2] In billions of deutsche mark in the columns for percentage changes.
[3] Figures differ slightly from those in Table A5 owing to different methods of accounting for social contributions of employers. In the data underlying this table, imputed values of certain social contributions are deducted from gross income from dependent labor in order to arrive at disposable income. In the data underlying Table A5, only actual contributions are deducted.
[4] Includes unrecorded tax payments, social contributions of the self-employed, net casualty insurance payments, repayments of current transactions to the state, and international private transfers.

Table A4. Federal Republic of Germany: Earnings, Wages, Employment, and Productivity

(Percentage change from previous year)

	1982	1983	1984	1985	1986	1987
Overall economy						
Gross earnings	1.8	1.8	3.4	3.7	5.0	3.8
Dependent employment	−1.9	−1.7	0.2	0.8	1.1	0.8
Hourly earnings	4.6	4.1	3.5	3.8	4.7	4.2
Hours per employee	−0.4	−0.3	−0.2	−0.7	−0.8	−1.2
Labor productivity[1]	1.1	3.8	3.3	2.0	2.3	2.2
Unit labor costs	3.4	0.4	0.3	1.7	2.4	1.9
Manufacturing and mining						
Gross earnings	4.4	2.5	0.3	3.3	3.1	3.3
Employment	−3.5	−3.9	−1.3	1.1	1.8	—
Hourly earnings	5.8	3.7	3.2	4.1	5.3	4.7
Hours per employee	2.2	2.9	−1.6	−1.2	−4.4	−1.3
Labor productivity[1]	2.0	4.6	4.6	3.7	1.7	1.7
Unit labor costs	3.8	−0.8	−1.3	0.3	3.6	3.0
GNP deflator	4.6	3.3	2.0	2.2	3.1	2.1

Sources: Deutsche Bundesbank, *Monthly Report, Supplement 4,* various issues; and Statistisches Bundesamt, *Wirtschaft und Statistik,* various issues.
[1] Output per hour.

APPENDIX I • STATISTICAL TABLES

Table A5. Federal Republic of Germany: Income and Consumption
(Percentage change from previous year)

	1980	1981	1982	1983	1984	1985	1986	1987
Real private consumption	1.3	−0.5	−1.3	1.7	1.5	1.7	4.3	3.1
Real household disposable income	1.9	0.2	−2.9	−0.3	2.4	1.4	4.7	3.5
Nominal disposable income[1]	7.4	6.5	2.3	3.0	4.8	3.6	4.5	3.7
Consumer price inflation	5.4	6.3	5.3	3.3	2.4	2.2	−0.2	0.2
Nominal interest rate[2]	8.5	10.4	9.0	7.9	7.8	6.9	5.9	5.8
Unemployment rate[3]	3.3	4.6	6.7	8.1	8.1	8.2	7.9	7.9
Personal saving ratio[4]	14.0	14.8	13.8	12.2	12.8	12.7	13.3	13.4

Sources: Deutsche Bundesbank, *Monthly Report, Supplement 4*, various issues; and Statistisches Bundesamt, *Volkswirtschaftliche Gesamtrechnungen*, various issues.
[1] Figures differ slightly from those in Table A3; see footnote 3 in that table.
[2] Nominal yield on outstanding public bonds of maturity longer than three years, in percent a year.
[3] In percent of total labor force; based on microcensus data.
[4] Household saving as a percent of household disposable income, Deutsche Bundesbank definition.

Table A6. Federal Republic of Germany: Balance of Payments Summary[1]
(In billions of deutsche mark)

	1983	1984	1985	1986	1987
Current account	13.5	27.9	48.4	85.0	80.8
Foreign trade[2]	42.1	54.0	73.4	112.6	117.7
Exports (f.o.b.)[2]	432.3	488.2	537.2	526.4	527.4
Imports (c.i.f.)[2]	−390.2	−434.3	−463.8	−413.7	−409.6
Supplementary items[3]	3.3	−1.1	−1.3	−1.5	−1.5
Services	−6.6	4.7	5.3	0.9	−7.1
Receipts	120.7	135.4	144.2	142.5	144.5
Expenditure	−127.3	−130.6	−138.9	−141.6	−151.6
Transfers	−25.2	−29.7	−28.9	−27.0	−28.3
Capital account					
(net capital exports: −)	−18.4	−37.5	−55.1	−80.1	−44.0
Long-term capital	−7.0	−19.8	−13.4	33.3	−23.6
German investment abroad	−36.5	−45.0	−61.7	−55.2	−62.5
Foreign investment in Germany	29.5	25.2	48.3	88.5	38.9
Short-term capital	−11.5	−17.7	−41.7	−113.4	−20.5
Financial institutions	1.8	0.1	−27.7	−59.0	−6.2
Enterprises	−8.9	−16.1	−14.1	−53.4	−11.6
Official	−4.3	−1.6	0.1	−1.0	−2.7
Balance of unclassifiable transactions (balancing item)	0.8	6.5	8.6	1.1	4.4
Change in the net external assets of the Deutsche Bundesbank[4]	−4.1	−3.1	1.8	6.0	41.2
Memorandum item					
Current account balance (*in percent of GNP*)	0.8	1.6	2.6	4.4	4.0

Source: Deutsche Bundesbank, *Monthly Report, Supplement 3*, various issues.
[1] Totals may differ from sums of components because of rounding.
[2] Excluding supplementary trade items.
[3] Including merchanting trade and warehouse transactions for account of residents, less the value of goods returned.
[4] At transactions values.

Table A7. Federal Republic of Germany: Merchandise Exports by Region

	1987 Levels[1]	1983	1984	1985	1986	1987
		(Percentage change from previous year)				
EC[2]						
Value	259.7	0.9	12.1	9.5	−1.0	2.9
Unit value	112.5	0.5	3.6	3.2	−3.7	−2.1
Volume	230.8	0.3	8.1	6.1	2.8	5.1
Other industrial[3]						
Value	191.3	5.0	22.2	14.3	2.5	0.3
Unit value	119.1	2.0	4.5	4.2	−2.2	−2.7
Volume	160.7	3.0	16.9	9.6	4.8	3.1
OPEC[4]						
Value	14.3	−17.0	−11.5	−9.8	−28.9	−20.2
Unit value	109.6	0.4	1.8	5.2	−4.9	−1.6
Volume	13.0	−17.3	−13.0	−14.3	−25.2	−18.9
Non-OPEC developing countries						
Value	38.3	2.8	11.6	4.3	−7.4	−0.2
Unit value	109.2	4.5	−0.7	5.6	−3.2	−4.0
Volume	35.1	−1.6	12.3	−1.2	−4.4	4.0
State-trading nations						
Value	23.0	10.2	3.2	19.5	−7.1	−11.2
Unit value	102.8	1.8	1.8	2.7	−5.5	−5.6
Volume	22.4	8.2	1.4	16.4	−1.7	−6.0

Source: Statistisches Bundesamt, *Aussenhandel*, various issues.
[1] Values are in billions of current deutsche mark; unit values are indices, with 1980 = 100; volumes are in billions of 1980 deutsche mark.
[2] Excluding Spain and Portugal.
[3] Including Spain and Portugal.
[4] Organization of Petroleum Exporting Countries.

Table A8. Federal Republic of Germany: Merchandise Exports by Product Category

	1987 Levels[1]	1983	1984	1985	1986	1987
		(Percentage change from previous year)				
Goods for final consumption						
Value	26.4	−1.0	12.9	5.1	−2.5	−3.4
Unit value	93.7	−3.3	5.4	0.9	−10.8	−7.7
Volume	28.2	2.3	7.1	4.2	9.3	4.6
Goods for use by producers						
Value	497.4	1.2	13.0	10.4	−2.0	0.4
Unit value	115.1	1.6	3.4	4.0	−2.8	−2.3
Volume	432.2	−0.4	9.3	6.1	0.8	2.8
Of which:						
Raw materials						
Value	6.6	−0.2	15.2	3.2	−15.0	−9.0
Unit value	104.4	−4.4	6.4	4.2	−5.9	−9.5
Volume	6.3	4.4	8.3	−0.9	−9.7	0.7
Semifinished goods						
Value	28.6	0.3	17.0	2.2	−25.3	−6.6
Unit value	89.1	−2.0	5.3	2.5	−18.9	−9.0
Volume	32.1	2.3	11.1	−0.3	−7.9	2.7
Intermediate goods						
Value	87.0	2.6	17.5	8.8	−5.5	−0.7
Unit value	102.7	−2.8	5.2	3.6	−6.2	−6.0
Volume	84.6	5.6	11.7	5.0	0.7	5.7
Final goods						
Value	375.2	1.0	11.3	12.0	1.8	1.4
Unit value	121.4	3.4	2.6	4.3	−0.3	−0.7
Volume	309.1	−2.4	8.5	7.3	2.1	2.0

Source: Statistisches Bundesamt, *Aussenhandel*, various issues.
[1] Values are in billions of current deutsche mark; unit values are indices, with 1980 = 100; volumes are in billions of 1980 deutsche mark.

Table A9. Federal Republic of Germany: Merchandise Imports by Region

	1987 Levels[1]	1983	1984	1985	1986	1987
		(Percentage change from previous year)				
EC[2]						
Value	204.7	5.7	8.7	8.4	−8.6	−0.6
Unit value	99.3	−0.5	4.3	3.0	−12.6	−4.8
Volume	206.1	6.3	4.1	5.2	4.6	4.3
Other industrial[3]						
Value	134.9	7.6	14.5	8.0	−4.7	1.1
Unit value	108.7	2.9	6.2	3.1	−10.7	−6.2
Volume	124.2	4.6	7.8	4.8	6.7	8.0
OPEC[4]						
Value	11.1	−16.9	—	−0.7	−51.2	−16.1
Unit value	60.0	−8.0	8.7	0.2	−54.9	−2.1
Volume	18.5	−9.6	−8.1	−0.9	8.3	−14.3
Non-OPEC developing countries						
Value	39.2	0.6	18.8	4.9	−11.3	−1.2
Unit value	87.2	—	10.7	—	−18.7	−12.2
Volume	44.9	0.5	7.4	4.9	9.2	12.4
State-trading countries						
Value	19.5	3.7	19.3	−0.5	−19.6	−8.0
Unit value	78.4	−1.7	5.6	3.6	−27.3	−14.1
Volume	24.8	5.5	13.0	−3.9	10.5	7.3

Source: Statistisches Bundesamt, *Aussenhandel*, various issues.
[1] Values are in billions of current deutsche mark; unit values are indices, with 1980 = 100; volumes are in billions of 1980 deutsche mark.
[2] Excluding Spain and Portugal.
[3] Including Spain and Portugal.
[4] Organization of Petroleum Exporting Countries.

Table A10. Federal Republic of Germany: Merchandise Imports by Product Category

	1987 Levels[1]	1983	1984	1985	1986	1987
		(Percentage change from previous year)				
Goods for final consumption						
Value	51.8	1.6	8.2	6.1	−5.5	−5.2
Unit value	96.3	−0.5	8.2	−1.7	−7.9	−9.1
Volume	53.8	2.2	—	7.9	2.6	4.1
Goods for use by producers						
Value	349.5	3.8	11.7	6.8	−11.9	−0.4
Unit value	97.8	−0.4	5.5	3.2	−17.3	−5.0
Volume	357.4	4.2	6.0	3.5	6.6	5.1
Of which:						
Raw materials						
Value	28.0	−12.0	11.6	−1.3	−46.3	−9.1
Unit value	64.9	−5.3	7.2	0.5	−47.4	−5.6
Volume	43.1	−7.1	4.1	−1.8	2.1	−3.1
Semifinished goods						
Value	51.0	4.8	11.3	6.6	−30.7	−15.1
Unit value	72.5	−1.5	7.6	2.6	−34.1	−16.1
Volume	70.3	6.4	3.4	3.9	5.2	1.1
Intermediate goods						
Value	59.5	6.9	12.5	7.8	−1.0	−2.1
Unit value	102.4	−1.6	6.2	4.3	−7.7	−6.1
Volume	58.1	8.6	5.9	3.4	7.2	4.1
Final goods						
Value	211.0	8.8	11.8	9.2	3.3	6.1
Unit value	113.5	2.9	3.8	4.1	−4.5	−3.1
Volume	185.9	5.7	7.7	4.9	8.2	10.1

Source: Statistisches Bundesamt, *Aussenhandel*, various issues.
[1] Values are in billions of deutsche mark; unit values are indices, with 1980 = 100; volumes are in billions of 1980 deutsche mark.

Table A11. Federal Republic of Germany: Regional Breakdown of Foreign Trade[1]

	Exports (f.o.b.)				Imports (c.i.f.)				Trade Balance		
	1985	1986	1987	1987	1985	1986	1987	1987	1985	1986	1987
	(Percent change from preceding period)			(Percent of share in total)	(Percent change from preceding period)			(Percent of share in total)	(In billions of deutsche mark)		
Industrial countries	11.5	0.5	1.8	85.6	8.2	−7.1	0.1	82.9	75.7	103.9	111.7
EC countries	9.7	0.1	3.9	52.7	8.5	−8.3	−0.2	52.6	31.6	51.4	62.3
Of which:											
Belgium-Luxembourg	8.7	0.6	4.5	7.4	1.0	0.5	−0.4	7.1	7.9	7.9	9.7
Denmark	17.7	3.4	−8.6	2.1	12.8	−4.5	0.1	1.9	3.8	4.6	3.5
France	4.3	−2.6	2.1	12.1	7.5	−4.5	0.8	11.6	14.7	15.2	16.1
Italy	11.0	2.6	7.4	8.7	8.7	2.5	2.9	9.6	4.6	4.8	6.9
Netherlands	9.8	−1.7	1.4	8.7	9.9	−18.0	−6.0	11.0	−12.0	−2.3	1.2
United Kingdom	13.3	−3.0	4.6	8.8	11.7	−19.9	−1.2	7.2	8.8	14.8	17.2
Other European countries	13.9	3.4	1.8	18.9	9.0	−4.4	1.9	16.0	27.6	33.8	34.4
Of which:											
Austria	12.6	2.6	1.0	5.4	11.8	6.7	5.6	4.2	12.0	11.7	11.1
Sweden	13.5	0.1	7.4	3.0	9.5	−8.2	−0.1	2.4	3.9	4.8	5.9
Switzerland	11.5	7.5	3.5	6.1	9.8	7.7	2.6	4.6	11.7	12.5	13.2
Non-European countries	14.7	−1.5	−5.4	14.0	6.4	−5.4	−1.0	14.3	16.5	18.7	15.0
Of which:											
Canada	26.7	−3.5	−9.9	0.9	−5.2	−12.3	−0.1	0.8	1.6	1.9	1.4
Japan	14.0	10.4	21.1	2.0	13.2	16.0	5.1	6.2	−12.8	−15.3	−14.7
South Africa	−24.8	−16.1	8.7	0.9	6.4	−8.9	−22.1	0.5	1.8	1.3	2.3
United States	18.6	−0.6	−9.6	9.5	4.0	−16.9	−4.7	6.3	23.2	28.4	24.4
Developing countries (excluding OPEC)[2]	4.1	−7.7	−0.5	7.2	4.9	−11.3	−1.3	9.5	−3.4	−1.5	−1.2
OPEC countries	−9.8	−28.9	−20.2	2.7	−0.7	−51.2	−16.1	2.7	−1.9	4.7	3.2
Of which:											
Iran	−25.8	−32.4	−13.4	0.5	−0.9	−38.5	−20.2	0.2	3.0	2.1	1.9
Saudi Arabia	−17.0	−35.6	−23.4	0.5	−22.7	−13.4	−43.3	0.3	3.1	1.5	1.5
State-trading countries	19.5	−7.1	−11.2	4.4	−0.5	−19.6	−8.0	4.8	1.6	4.7	3.5
Of which:											
Hungary	11.6	−2.2	−3.5	0.5	8.9	−7.1	3.5	0.5	0.8	0.9	0.7
Poland	20.2	−13.6	−2.4	0.5	11.4	−15.7	−4.6	0.6	−0.2	−0.1	−0.1
U.S.S.R.	−2.2	−11.0	−16.3	1.5	−5.3	−31.8	−21.9	1.8	−3.1	0.1	0.6
Total[3]	10.0	−2.0	0.2	100.0	6.8	−10.8	−1.0	100.0	73.4	112.6	117.7

Source: Deutsche Bundesbank, *Monthly Report, Supplement 3*, various issues.
[1] Excluding supplementary trade items.
[2] Organization of Petroleum Exporting Countries.
[3] Including trade with unspecified countries accounting for 0.2 percent of exports, 0.1 percent of imports, and for a trade surplus of DM 0.6 billion in 1987.

Table A12. Federal Republic of Germany: Service Account Receipts and Payments

(In billions of deutsche mark)[1]

	1983	1984	1985	1986	1987
Tourism	−23.2	−23.1	−23.7	−25.7	−28.5
Receipts	10.8	12.2	14.0	13.7	13.9
Payments	33.9	35.4	37.7	39.4	42.4
Transportation	9.3	9.7	10.9	9.1	8.1
Investment income	4.2	10.3	9.5	8.7	6.7
Receipts	34.5	41.6	44.0	46.2	50.4
Payments	30.3	31.3	34.5	37.4	43.7
Government payments[2]	14.8	17.7	20.3	19.6	19.2
Other	−11.5	−9.6	−11.3	−11.0	−12.9
Total (net)	−6.6	4.7	5.3	0.9	−7.1

Source: Deutsche Bundesbank, *Monthly Report, Supplement 3*, various issues.

[1] Totals may differ from sums of components because of rounding.
[2] Balance on payments among official institutions both at home and abroad, which are not listed elsewhere in the balance of payments accounts; reflects mostly receipts of the German Government for deliveries of goods and services to foreign military posts in Germany.

Table A13. Federal Republic of Germany: Real Value Added by Sector

	1987 Value added at current prices		1961–70	1971–80	1981–85	1984	1985	1986	1987
			Annual averages			At 1980 prices			
	(In billions of deutsche mark)	(In percent of total)	(Percentage change from previous year)						
Manufacturing	664,700	35.1	5.7	2.1	0.7	2.7	3.5	2.5	0.6
Transport and communication	110,170	5.8	4.6	3.8	2.5	4.2	3.6	2.4	2.8
Agriculture, forestry, and fishing	29,360	1.6	1.5	0.7	2.6	8.7	−5.4	7.7	−8.9
Tradables sectors	804,230	42.5	5.3	2.3	1.0	3.3	3.0	2.7	0.4
Services (including finance, government, nonprofit)	811,900	42.9	4.3	4.0	2.6	3.1	3.3	3.3	3.2
Trade	174,180	9.2	4.9	2.5	1.4	4.1	1.4	2.5	1.3
Construction	101,640	5.4	3.4	1.1	−2.3	−0.1	−3.8	2.0	0.1
Nontradables sectors	1,087,720	57.5	4.2	3.3	1.8	2.9	2.2	3.0	2.6
Total[1]	1,891,950	100.0	4.7	2.8	1.5	3.1	2.5	2.9	1.6

Source: Fund staff calculations based on Statistisches Bundesamt, *Volkswirtschaftliche Gesamtrechnungen*, various issues.

[1] Energy and mining are not included, and no correction is made for imputed output of bank services.

Table A14. Federal Republic of Germany: Value-Added Deflators by Sector

	1987		1961–70	1971–80	1981–85	1984	1985	1986	1987
	Value added at current prices		Annual averages			Value-added deflators			
	(In billions of deutsche mark)	(In percent of total)	(Percentage change from previous year)						
Manufacturing	664,700	35.1	2.8	4.2	3.2	1.5	3.1	7.3	2.9
Transport and communication	110,170	5.8	1.7	4.9	1.9	1.9	1.2	0.4	−0.7
Agriculture, forestry, and fishing	29,360	1.6	1.2	3.0	−1.1	−1.0	−2.6	−0.9	−5.5
Tradables sectors	804,230	42.5	2.5	4.2	2.8	1.3	2.5	5.7	2.2
Services (including finance, government, nonprofit)	811,900	42.9	6.3	6.8	3.7	2.1	1.7	2.0	2.5
Trade	174,180	9.2	2.1	5.3	2.6	1.2	0.8	−2.9	1.4
Construction	101,640	5.4	5.7	5.7	1.3	0.7	−0.5	2.7	3.3
Nontradables sectors	1,087,720	57.5	5.1	6.3	3.3	1.8	1.4	1.2	2.4
Total[1]	1,891,950	100.0	3.8	5.3	3.1	1.6	1.9	3.1	2.3

Source: Fund staff calculations based on Statistisches Bundesamt, *Volkswirtschaftliche Gesamtrechnungen*, various issues.

[1] Energy and mining are not included, and no correction is made for imputed output of bank services.

APPENDIX I • STATISTICAL TABLES

Table A15. Federal Republic of Germany: Growth of the Main Monetary Aggregates[1]
(Period averages, seasonally adjusted)[2]

	Dec. 1987 Level	1982	1983	1984	1985	1986	1987	1987 I	1987 II	1987 III	1987 IV	1988 I	1988 II	Memorandum Items Growth from IV 1985–IV 1986	Memorandum Items Growth from IV 1986–IV 1987
	(In billions of deutsche mark)	*(Percentage change from preceding period at annual rate)*													
Central bank money stock[3]	238.7	4.9	7.3	4.8	4.6	6.4	8.1	7.6	8.4	7.1	9.1	8.5	7.7	7.7	8.1
Narrow money, M1	363.8	3.6	10.3	3.3	4.3	8.5	9.0	7.2	12.2	10.0	4.4	13.1	12.8	8.6	8.4
Of which: Currency outside banks	122.1	3.0	8.7	5.4	3.6	7.0	9.2	9.8	10.7	9.7	10.3	15.6	9.8	7.7	10.2
Demand deposits	241.7	3.9	11.0	2.2	4.7	9.3	8.8	6.0	13.0	10.1	1.7	11.9	14.3	9.1	7.6
Broad money, M2	619.7	6.8	2.9	3.4	4.4	5.4	6.7	8.5	4.8	3.1	6.3	8.6	6.0	6.9	5.7
Of which: Time deposits with maturities of less than four years	256.0	10.4	−5.2	3.6	4.4	1.4	3.6	10.4	−4.6	−6.1	9.0	2.3	−3.4	4.6	1.9
M3 (M2 plus savings deposits at statutory notice)	1,080.1	6.5	6.6	3.8	4.9	5.9	7.1	8.3	5.7	4.8	5.5	8.2	7.0	7.2	6.1
Of which: Savings deposits at statutory notice	460.4	6.2	12.4	4.5	5.5	6.5	7.6	8.0	6.9	7.1	4.4	7.7	8.4	7.6	6.6
Monetary capital formation	1,341.1	7.4	6.0	8.8	7.4	7.3	6.3	6.8	6.1	5.9	4.7	1.8	1.8	7.1	5.9
Of which: Savings deposits and savings bonds	424.0	5.7	4.5	6.7	6.6	7.4	6.5	7.5	5.9	4.8	2.9	−0.3	−2.1	7.7	5.2
Time deposits with maturities of four years or more	407.1	4.2	8.1	10.7	11.2	11.4	11.2	12.4	11.5	12.5	11.3	13.3	11.8	10.5	11.9
Bank bonds outstanding[4]	377.4	10.8	5.7	10.1	4.8	2.6	0.8	0.5	0.6	−0.4	−0.9	−8.2	−4.5	2.5	—
Bank loans outstanding to domestic nonbanks	2,180.7	7.4	6.3	6.3	6.1	4.3	3.5	3.0	2.5	5.9	5.5	4.9	5.6	3.9	4.2
Public authorities	494.7	13.4	7.8	3.1	5.9	1.7	3.1	1.8	3.9	8.2	10.2	10.4	6.9	1.0	6.0
Private nonbanks	1,686.0	5.7	5.8	7.3	6.2	5.1	3.7	3.3	2.1	5.3	4.2	3.4	5.1	4.7	3.7
Short-term[5]	321.4	6.3	3.2	7.3	4.6	−0.1	−4.8	−4.4	−9.4	−2.7	−1.1	−0.9	3.6	−1.8	−4.4
Long-term	1,364.6	5.5	6.6	7.3	6.6	6.6	6.0	5.4	5.2	7.3	5.5	4.4	5.5	6.6	5.9

Source: Deutsche Bundesbank, *Monthly Report*, *Supplement 4*, various issues.
[1] From December 1985, the published data are not comparable with earlier data because of extended coverage; to construct this table, the pre-December 1985 and post-December 1985 series have been spliced together based on the December 1985 values under the old and new coverage.
[2] For series other than central bank money, data are averages of end-of-month levels.
[3] Currency held by nonbanks plus minimum reserve requirements on domestic bank liabilities (at constant reserve ratios of January 1974).
[4] Other than bank holdings.
[5] Up to one year.

Table A16. Federal Republic of Germany: Indicators of the Internationalization of the German Capital Markets[1]

	Nonresident Purchases of Domestic Bonds	Resident Purchases of Foreign Bonds[2]	Resident Purchases of Foreign Currency Bonds	Nonresident Purchases of Domestic Equity	Resident Purchases of Foreign Equity
	(In percent of issues)	(In percent of total resident acquisition of bonds)	(In percent of total resident acquisition of bonds)	(In percent of issues)	(In percent of total resident acquisition of equity)
1975–79[3]	2.0	6.2	2.5	35.9	47.5
1980–83[3]	4.4	10.5	7.3	27.7	51.6
1984–87[3]	43.1	31.5	24.3	54.9	62.3
1980	0.7	14.0	2.7	15.8	37.9
1981	−2.2	8.3	5.5	55.9	65.6
1982	3.1	13.5	12.7	7.7	37.5
1983	12.6	7.1	6.6	33.8	63.3
1984	19.4	21.5	19.2	63.6	71.3
1985	41.4	38.1	28.1	65.5	66.3
1986	67.5	36.5	21.3	92.6	92.9
1987	39.7	31.8	27.5	−11.4	27.6

Sources: Deutsche Bundesbank, *Monthly Report* and *Monthly Report, Supplement 2*, various issues.
[1] All purchase and issue data are on a net basis.
[2] Includes acquisition of foreign deutsche mark bonds.
[3] Calculations are based on the sum of flows for the entire period.

APPENDIX I • STATISTICAL TABLES

Table A17. Federal Republic of Germany: Exchange Rates
(Percentage change from previous period)[1]

						1987				1988		Memorandum Items	
	1983	1984	1985	1986	1987	I	II	III	IV	I	II	End 1986–end 1987[2]	End 1987–Sept. 1988[3]
												(Percentage change)	
Effective exchange rates													
Nominal													
All countries[4]	3.9	−1.1	—	8.8	6.0	2.6	−0.7	−0.2	1.5	−0.2	−1.1	3.5	−3.1
EMS[5]	6.8	2.5	1.0	3.6	3.2	1.2	0.4	0.2	0.8	0.2	0.2	2.5	0.5
Real													
All countries (ULC based)[6]	1.5	−2.4	−0.6	8.7	6.0	2.5	−0.1	−1.1	2.1	1.3	−2.7	4.5	−3.2
All countries (CPI based)[7]	0.6	−4.4	−2.7	5.7	3.3	2.1	−1.1	−0.7	0.9	−0.5	−1.6	1.1	...
EMS (ULC based)[8]	2.1	−0.5	−0.1	5.3	4.7	1.3	1.2	−0.5	1.9	2.1	−1.3	3.9	...
EMS (CPI based)[8]	0.9	−2.4	−2.5	0.5	0.9	1.2	—	−0.5	0.3	0.1	−0.1	0.9	...
Bilateral exchange rates													
U.S. dollar	−4.9	−10.2	−2.9	35.1	20.5	9.1	1.9	−1.8	8.0	1.6	−1.8	22.7	−15.9
French franc	10.2	3.0	−0.6	4.6	4.7	1.7	0.2	—	1.2	0.2	0.1	2.3	0.3
Pound sterling	9.6	2.1	0.2	18.9	8.3	1.2	−4.4	−0.2	−0.5	−0.7	−4.2	−3.2	−6.6
Japanese yen	−9.3	−10.1	−3.0	−4.5	3.9	4.2	−5.1	1.2	−0.5	−4.0	−3.6	−6.8	−6.8

Sources: Deutsche Bundesbank, *Monthly Report*, various issues; and International Monetary Fund, Data Fund.
[1] Positive figures indicate appreciation of the deutsche mark.
[2] For effective rates (real and nominal) against all countries, calculations are based on December average data; for real effective rates, against exchange rate mechanism (ERM) countries, the calculation is based on fourth quarter data; for bilateral rates and the nominal effective rate against ERM countries, end-of-year data are used.
[3] Based on monthly average data for effective exchange rates and end-of-month data for bilateral exchange rates.
[4] As used for the Information Notice System (INS).
[5] As calculated by Bundesbank for countries participating in the ERM.
[6] As used for the INS based on normalized unit labor costs (ULC) in manufacturing and exchange rates vis-à-vis 16 other industrial countries.
[7] As calculated for the INS. The consumer price based index uses a wider range of countries in its weighting scheme—35 other countries, compared with 16 other countries used for the unit labor cost based calculation.
[8] Staff calculations for countries participating in the ERM. The ULC measure is based on normalized unit labor costs in manufacturing.

Statistical Tables

Table A18. Federal Republic of Germany: Changes in Central Bank Money and Its Determinants[1]

(In billions of deutsche mark)

	Purchases(+) or Sales(−) of Foreign Exchange[2]	Net Credit to the Government[3]	Change in Reserve Requirements[4]	Change in Redis-count Quotas[5]	Open Market Operations Under Repurchase Agreements	Other Open Market Operations	Foreign Exchange Swaps, etc.[6]	Shifts of Federal Balance to the Money Market[7]	Change in Use of Redis-count Quotas[8]	Lombard or Special Lombard Loans	Other	Central Bank Money (increase (−))	Currency in Circulation (increase (−))	Minimum Reserves on Domestic Liabilities (increase (−))[9]
1980	−24.6	0.3	10.0	12.1	6.0	4.5	4.6	−0.1	−1.4	2.6	−7.5	−6.5	−4.2	−2.3
1981	−3.1	1.1	3.6	5.1	4.4	−0.1	−0.7	0.2	1.3	−2.5	−6.6	−2.7	0.2	−2.9
1982	1.7	−4.1	5.2	7.7	−1.4	1.5	0.3	1.3	−3.5	0.1	−1.4	−7.5	−4.3	−3.1
1983	−2.0	1.7	—	−0.7	6.6	2.4	−1.9	−1.5	3.3	1.0	1.1	−10.1	−7.3	−2.8
1984	−3.9	1.3	—	7.8	7.7	−3.9	—	—	−1.0	0.3	−1.1	−7.1	−4.6	−2.6
1985	−0.7	−4.2	—	3.3	16.5	−0.6	0.2	1.2	−3.1	−5.0	−1.0	−6.6	−3.9	−2.7
1986	8.7	−0.2	7.2	−5.6	−9.5	2.2	0.3	0.4	4.2	0.6	4.9	−13.1	−8.6	−4.5
1987	38.7	1.8	−5.4	−7.6	−5.5	−1.4	−0.3	−1.6	−0.1	−0.9	−2.3	−15.5	−11.5	−4.1
1987														
Quarter I	15.8	−3.7	−5.4	−7.2	1.8	−1.6	−0.5	—	0.9	−0.6	−3.4	3.9	3.6	0.3
Quarter II	4.8	2.6	—	−0.3	−5.4	−0.1	—	−1.7	−0.8	−0.1	6.1	−5.1	−4.3	−0.8
Quarter III	−2.6	−2.2	—	—	8.2	1.0	—	1.9	0.8	−0.2	−3.5	−3.2	−2.4	−0.8
Quarter IV	20.7	5.1	—	−0.1	−10.0	−0.7	0.2	−1.8	−0.8	—	−1.4	−11.1	−8.3	−2.8
1988														
Quarter I	0.2	−7.8	—	−5.0	9.2	—	−0.2	1.5	1.4	−0.1	−1.2	2.0	1.7	0.3
Quarter II	−5.8	5.9	—	−0.3	2.8	0.4	—	0.6	0.1	0.6	−0.5	−3.9	−3.4	−0.4

Source: Deutsche Bundesbank, *Monthly Report*, various issues.

[1] Period totals calculated on the basis of the daily averages of the months. All but the last three columns show factors giving rise to changes in central bank money. A positive sign indicates a factor increasing central bank money in the period. The last three columns show the change in central bank money and how it is distributed between currency and minimum reserves. For each period, the change in central bank money is calculated at the current reserve rates but excludes the effects of changes in the period concerned. The definition of central bank money in this table, therefore, does not correspond to the concept used in the setting of the Bundesbank's monetary target up to 1987, since it does not adjust the change in required minimum reserves in any period for any changes in reserve requirement ratios that occurred between 1974 and the beginning of the period concerned. The quarterly data are also not seasonally adjusted. Thus the central bank money data in this table are not on the same basis as those used to construct the growth rates in Table A1, which do correspond to the definition used in constructing the monetary target and which are seasonally adjusted.
[2] Excluding foreign exchange swaps and foreign exchange transactions under repurchase agreement.
[3] Changes in net deposits of central and regional authorities at the Bundesbank. These deposits are defined to include federal funds shifted temporarily to the banks (under Section 17 of the Bundesbank Act).
[4] Domestic liabilities. Lowering of required reserve ratios (+).
[5] Rediscount quotas, including limits for money market papers eligible for purchase by the Bundesbank.
[6] Foreign exchange swaps and foreign exchange transactions under repurchase agreement.
[7] Under Section 17 of the Bundesbank Act.
[8] Decline in unutilized rediscount facilities.
[9] At current reserve ratios, but excluding the effects of changes in minimum reserves during the period concerned.

Table A19. Federal Republic of Germany: Interest Rates

(In percent a year)

	Bundesbank Rates[1]			Money Market Rate[2]	Government Bond Yields[2]		Memorandum Items[2]	
	Discount rate	Securities repurchase rate[3]	Lombard rate[4]	Three-month rate	Average yield[5]	Long-term bonds[6]	Eurodollar 3-month LIBOR	U.S. Government 10-year bonds
1979 December	6.0	...[7]	7.0	9.6	7.9	7.9	14.7	10.4
1980 December	7.5	9.5	9.0	10.2	8.9	8.8	19.6	12.8
1981 December	7.5	10.3	10.5	10.8	9.7	9.7	13.4	13.7
1982 December	5.0	5.9	6.0	6.6	7.9	8.0	9.6	10.5
1983 December	4.0	6.0	5.5	6.5	8.2	8.4	10.2	11.8
1984 December	4.5	5.5	5.5	5.8	7.0	7.2	9.1	11.5
1985 December	4.0	4.6	5.5	4.8	6.5	6.6	8.1	9.3
1986 December	3.5	4.6	5.5	4.8	5.9	6.1	6.4	7.1
1987 December	2.5	3.3	4.5	3.7	6.0	6.5	8.0	9.0
1987 January	3.0	4.6	5.0	4.5	5.8	6.0	6.2	7.1
February	3.0	3.8	5.0	4.0	5.7	6.1	6.4	7.3
March	3.0	3.8	5.0	4.0	5.6	6.0	6.5	7.3
April	3.0	3.8	5.0	3.9	5.5	5.8	6.9	8.0
May	3.0	3.6	5.0	3.8	5.4	5.7	7.4	8.6
June	3.0	3.6	5.0	3.7	5.6	6.0	7.2	8.4
July	3.0	3.6	5.0	3.8	5.8	6.3	7.0	8.5
August	3.0	3.6	5.0	4.0	6.0	6.5	7.0	8.8
September	3.0	3.6	5.0	4.0	6.2	6.7	7.6	9.4
October	3.0	3.8	5.0	4.7	6.5	6.9	8.4	9.5
November	3.0	3.4	4.5	3.9	6.0	6.4	7.5	8.9
December	2.5	3.3	4.5	3.7	6.0	6.5	8.0	9.0
1988 January	2.5	3.3	4.5	3.4	6.0	6.5	7.3	8.7
February	2.5	3.3	4.5	3.3	5.8	6.3	6.8	8.2
March	2.5	3.3	4.5	3.4	5.7	6.2	6.9	8.4
April	2.5	3.3	4.5	3.4	5.8	6.3	7.2	8.7
May	2.5	3.3	4.5	3.5	6.1	6.6	7.5	9.1
June	2.5	3.3	4.5	3.9	6.1	6.6	7.7	8.9
July	3.0	3.8	5.0	4.9	6.4	6.8	8.2	9.1
August	3.5	4.3	5.0	5.3	6.5	6.8	8.6	9.3
September	3.5	4.3[8]	5.0	5.0	6.3	6.6	8.4	9.0
Memorandum items[9]								
1985 average	4.3	5.2	5.8	5.4	6.8	7.0	8.4	10.6
1986 average	3.6	4.4	5.5	4.6	5.9	6.1	6.9	7.7
1987 average	3.0	3.7	4.9	4.0	5.8	6.2	7.2	8.4

Source: Deutsche Bundesbank, *Monthly Report*, various issues.

[1] End of period for discount rate and Lombard rate; period average for rate on securities repurchase transactions.
[2] Period average.
[3] Unweighted average of rates during period for securities repurchase agreements.
[4] Interest rate on special Lombard loans during suspension of the general Lombard facility, effective February 20, 1981 to May 6, 1982.
[5] Average yield on public authority bonds with remaining maturities of 3 years or more.
[6] Average yield on public authority bonds with remaining maturities of 9–10 years.
[7] No transactions.
[8] Based on rates for fixed rate tenders only.
[9] Averages of end-of-month rates for discount rate and Lombard rate.

Table A20. Federal Republic of Germany: Sales and Purchases of Securities[1]

(In billions of deutsche mark)

	1983	1984	1985	1986	1987	1987 I	1987 II	1987 III	1987 IV	1988 I	1988 II
Bonds (total sales = total purchases)	**91.3**	**86.8**	**103.5**	**103.8**	**113.0**	**49.7**	**22.1**	**28.6**	**12.7**	**29.8**	**17.8**
Sales											
Domestic bonds	85.5	71.1	76.1	87.5	88.2	42.5	13.9	19.9	11.9	16.3	4.0
Bank bonds	51.7	34.6	33.0	29.5	28.4	15.3	5.3	6.5	1.3	−1.8	−2.5
Public authority bond	34.4	36.7	42.7	57.8	59.8	27.2	8.7	13.3	10.6	18.2	6.5
Industrial bonds	−0.6	−0.2	0.3	0.2	—	−0.1	−0.1	0.2	—	—	—
Foreign bonds[2]	5.7	15.7	27.5	16.3	24.8	7.2	8.1	8.7	0.8	13.5	13.9
Purchases											
Domestic purchases	80.5	73.0	72.1	44.7	78.0	26.5	10.1	29.0	12.4	27.4	19.3
Banks	35.2	26.4	32.7	31.3	44.3	12.5	7.4	12.7	11.7	10.3	8.4
Deutsche Bundesbank	2.4	−3.5	−0.2	1.1	−0.7	−0.2	−0.2	−0.1	−0.1	—	0.1
Nonbanks	42.9	50.0	39.5	12.4	34.4	14.2	2.9	16.5	0.8	17.1	10.8
Foreign purchases[3]	10.8	13.8	31.5	59.1	35.0	23.2	12.0	−0.5	0.3	2.4	−1.4
Shares (total sales = total purchases)	**15.6**	**11.9**	**18.5**	**32.2**	**16.9**	**1.9**	**5.6**	**2.2**	**7.2**	**5.4**	**5.4**
Sales											
Domestic shares	7.3	6.3	11.0	16.4	11.9	1.6	4.7	2.0	3.6	1.1	0.8
Foreign shares[4]	8.3	5.7	7.5	15.8	5.0	0.4	0.9	0.1	3.6	4.4	4.6
Purchases											
Domestic purchases	13.1	8.0	11.3	17.1	18.3	0.5	4.2	−0.8	14.4	6.9	7.5
Banks	0.7	1.5	2.5	5.9	3.8	0.9	2.0	1.3	−0.4	1.3	−0.2
Nonbanks	12.4	6.4	8.8	11.2	14.5	−0.5	2.2	−2.1	14.8	5.6	7.7
Foreign purchases[5]	2.5	4.0	7.2	15.2	−1.4	1.5	1.4	2.9	−7.2	−1.5	−2.1
Memorandum items											
Overall balance of security transactions with abroad (capital imports (+))	−0.8	−3.6	3.8	42.1	3.7	17.0	4.4	−6.4	−11.3	−16.9	−22.0
Of which: Bonds	*5.1*	*−1.9*	*4.0*	*42.7*	*10.1*	*15.9*	*3.9*	*−9.2*	*−0.5*	*−11.1*	*−15.3*
Shares	*−5.9*	*−1.7*	*−0.2*	*−0.7*	*−6.4*	*1.1*	*0.5*	*2.8*	*−10.8*	*−5.8*	*−6.6*

Source: Deutsche Bundesbank, *Monthly Report,* various issues.

[1] Discrepancies in totals are due to rounding.
[2] Net purchases (+) or sales (−) of foreign bonds by residents.
[3] Net purchases (+) or sales (−) of domestic bonds by nonresidents.
[4] Net purchases (+) or sales (−) of foreign shares (including direct investment and investment funds) by residents.
[5] Net purchases (+) or sales (−) of domestic shares (including direct investment and investment funds) by nonresidents.

APPENDIX I • STATISTICAL TABLES

Table A21. Federal Republic of Germany: Long-Term Capital in the Balance of Payments Accounts[1]

(In billions of deutsche mark)

	1982	1983	1984	1985	1986	1987	1987 I	1987 II	1987 III	1987 IV	1988 I	1988 II
German investment abroad	**−28.3**	**−36.5**	**−45.0**	**−61.7**	**−55.2**	**−62.5**	**−14.7**	**−14.8**	**−18.2**	**−14.7**	**−22.6**	**−24.1**
Direct investment	−6.0	−8.1	−12.5	−14.1	−20.3	−16.5	−4.2	−4.0	−3.4	−4.9	−3.4	−4.8
Advances and loans of enterprises	−1.1	−0.8	−1.7	0.3	0.2	−0.7	−0.3	—	—	−0.4	−0.3	−0.1
Portfolio investment	−11.4	−10.4	−15.7	−31.5	−21.6	−24.8	−6.1	−7.5	−9.3	−1.9	−16.6	−17.1
Of which:												
Foreign currency bonds	*−10.3*	*−5.3*	*−14.0*	*−20.2*	*−9.5*	*−21.5*	*−5.0*	*−7.2*	*−7.6*	*−1.7*	*−9.3*	*−11.3*
Deutsche mark bonds	*−0.7*	*−0.5*	*−1.7*	*−7.2*	*−6.8*	*−3.4*	*−2.2*	*−0.9*	*−1.1*	*0.9*	*−4.2*	*−2.6*
Shares	*−0.4*	*−4.6*	*—*	*−4.1*	*−5.3*	*0.1*	*1.2*	*0.6*	*−0.7*	*−1.1*	*−3.1*	*−3.2*
Advances and loans of banks	−3.7	−8.4	−6.8	−8.4	−6.3	−13.8	−3.1	−1.3	−4.6	−4.8	−1.2	−0.6
Official[2]	−4.7	−6.8	−6.9	−6.3	−5.4	−5.0	−0.7	−1.5	−0.5	−2.3	−0.8	−1.2
Real estate investment	−1.3	−1.3	−1.0	−1.0	−1.0	−1.0	−0.2	−0.3	−0.2	−0.3	−0.2	−0.3
Other	−0.2	−0.7	−0.4	−0.7	−0.7	−0.7	−0.1	−0.3	−0.2	−0.1	−0.1	−0.1
Foreign investment in Germany	**14.2**	**29.5**	**25.2**	**48.3**	**88.5**	**38.9**	**31.8**	**16.4**	**—**	**−9.3**	**−1.3**	**−2.6**
Direct investment	2.0	4.5	1.6	1.8	2.2	3.5	1.5	0.3	1.4	0.3	0.8	−2.6
Advances and loans to enterprises	3.9	1.3	0.5	—	0.7	2.1	1.6	1.1	−0.3	−0.3	0.4	3.1
Portfolio investment	12.2	25.5	21.7	36.3	69.0	20.9	22.5	12.0	−2.9	−10.7	−2.4	−4.1
Of which:												
Bonds	*2.3*	*10.8*	*13.8*	*31.5*	*59.1*	*35.0*	*23.2*	*12.0*	*−0.5*	*0.3*	*2.4*	*−1.4*
Official borrowers' notes	*9.4*	*11.9*	*4.3*	*−2.0*	*−5.1*	*−12.3*	*−2.4*	*−2.2*	*−3.9*	*−3.8*	*−3.2*	*−2.7*
Advances and loans to banks	−3.9	−1.6	1.5	10.3	16.6	12.6	6.3	3.1	1.7	1.5	−0.2	1.1
Other	−0.1	−0.2	—	−0.1	−0.1	−0.1	—	—	—	—	—	—
Balance on long-term capital account	**−14.2**	**−7.0**	**−19.8**	**−13.4**	**33.3**	**−23.6**	**17.1**	**1.6**	**−18.3**	**−23.9**	**−23.9**	**−26.7**

Source: Deutsche Bundesbank, *Monthly Report, Supplement 3*, various issues.
[1] Outflows of funds have a negative sign. Totals may differ from sums of components because of rounding.
[2] Includes share contributions to international organizations.

Table A22. Federal Republic of Germany: Short-Term Capital in the Balance of Payments Accounts[1]

(In billions of deutsche mark)

	1982	1983	1984	1985	1986	1987	1987 I	1987 II	1987 III	1987 IV	1988 I	1988 II
Banks	8.1	1.8	0.1	−27.7	−59.0	−6.2	−11.1	−12.6	9.3	8.2	13.9	−5.6
Assets	4.3	5.3	−17.8	−33.4	−65.8	−15.4	−7.1	−17.4	5.7	3.4	10.4	−7.3
Liabilities	3.8	−3.6	17.8	5.7	6.8	9.3	−4.0	4.8	3.6	4.8	3.5	1.7
Enterprises and individuals	3.2	−8.9	−16.1	−14.1	−53.4	−11.6	−6.1	−3.9	−8.4	6.8	−11.2	−2.1
Financial credits	5.1	−3.1	−7.4	−10.7	−48.3	−13.5	−7.5	−4.0	−5.4	3.5	−7.2	0.4
Assets	−0.9	−2.6	−10.0	−12.1	−35.1	−10.1	−7.0	−1.9	−6.3	5.1	−10.2	−0.1
Liabilities	6.0	−0.4	2.6	1.5	−13.2	−3.4	−0.6	−2.1	0.9	−1.7	3.0	0.6
Trade credits	−1.9	−5.9	−8.7	−3.5	−5.2	1.9	1.4	0.1	−2.9	3.3	−4.0	−2.6
Assets	−4.6	−7.7	−11.7	−3.9	−0.5	1.2	1.6	−0.2	−2.3	2.1	−4.9	−5.1
Liabilities	2.8	1.8	3.0	0.4	−4.6	0.7	−0.1	0.3	−0.6	1.2	0.9	2.5
Official	−0.3	−4.3	−1.6	0.1	−1.0	−2.7	−2.0	0.1	0.1	−0.9	−2.0	0.2
Assets	−0.5	−0.9	−0.4	0.6	−0.2	−3.3	−1.7	−0.8	0.3	−1.1	−1.8	−0.2
Liabilities	0.2	−3.4	−1.2	−0.5	−0.7	0.6	−0.3	0.9	−0.2	0.2	−0.2	0.5
Balance of short-term capital transactions	11.0	−11.5	−17.7	−41.7	−113.4	−20.5	−19.2	−16.4	1.1	14.1	0.7	−7.6
Memorandum items												
Balancing item of the balance of payments	−6.2	0.8	6.5	8.6	1.1	4.4	−4.5	0.8	1.4	6.7	5.8	−1.4
Short-term capital transaction, including balancing item in the balance of payments	4.8	−10.6	−11.2	−33.2	−112.3	−16.0	−23.7	−15.5	2.5	20.8	6.5	−9.0

Source: Deutsche Bundesbank, *Monthly Report, Supplement 3,* various issues.
[1] Outflows have a negative sign. Totals may differ from sums of components because of rounding.

Table A23. Federal Republic of Germany: Distribution of the Ownership of Shares, December 1987

	Share Holdings[1] (In billions of deutsche mark)	Percentage Distribution
Households	118.4	19
Nonfinancial enterprises	253.2	40
Government	44.7	7
Banks	62.3	10
Insurance enterprises	73.2	12
Other domestic	0.3	—
Foreign holdings	81.3	13
Total	633.4	100

Source: Deutsche Bundesbank, *Monthly Report* (May 1988).
[1] Valued at market prices; includes domestic holdings of foreign shares.

APPENDIX I • STATISTICAL TABLES

A24. Federal Republic of Germany: Selected Indicators of the Size of the General Government[1]
(In percent of GNP)

	1970–74	1975–79	1980	1981	1982	1983	1984	1985	1986	1987	1988[2]
Revenue	41.5	45.2	45.7	45.9	46.6	46.1	46.1	46.3	45.5	45.1	44.5
Of which:											
Indirect taxes	13.1	12.8	13.0	12.8	12.6	12.8	12.8	12.5	12.1	12.2	12.0
Direct taxes	11.7	12.8	12.6	12.2	12.1	11.9	12.0	12.4	12.2	12.1	11.9
Social security contributions	13.8	16.5	16.7	17.4	17.8	17.3	17.2	17.4	17.3	17.3	17.5
Other	2.8	3.1	3.3	3.5	4.0	4.1	4.0	4.0	3.9	3.4	3.1
Expenditure	41.6	48.5	48.6	49.6	49.8	48.6	48.0	47.4	46.7	46.8	46.8
Of which:											
Public consumption	17.4	19.8	20.1	20.6	20.4	20.0	19.8	19.8	19.6	19.6	19.5
Public gross investment	4.2	3.5	3.6	3.2	2.8	2.5	2.4	2.3	2.4	2.4	2.3
Interest payments	1.1	1.6	1.9	2.3	2.8	3.0	3.0	3.0	2.9	2.9	2.9
Transfer payments	19.0	23.7	23.1	23.5	23.8	23.1	22.8	22.3	21.8	21.9	22.1
Of which:											
Social transfers to households	13.4	17.1	16.5	17.2	17.6	17.0	16.3	16.0	15.7	15.9	15.9
Other transfers to households	1.2	1.2	1.0	1.0	1.0	0.9	0.8	0.8	0.8	0.8	0.8
Subsidies	1.9	2.1	2.1	1.9	1.8	1.9	2.0	2.0	2.1	2.2	2.1
Other transfers to enterprises	1.4	1.7	1.8	1.7	1.7	1.6	1.8	1.6	1.5	1.4	1.4
Transfers to abroad	1.1	1.5	1.7	1.7	1.7	1.7	1.9	1.8	1.7	1.7	1.9
Financial balance[3]	−0.1	−3.3	−2.9	−3.7	−3.3	−2.5	−1.9	−1.1	−1.2	−1.7	−2.3
Savings[3]	5.2	1.4	1.8	0.5	0.4	0.7	1.3	2.0	1.8	1.2	...
Employment[4]	12.2	14.4	14.9	15.2	15.6	15.9	16.0	16.1	16.1	16.2	...

Sources: Data provided by the Ministry of Finance; Statistisches Bundesamt, *Wirtschaft und Statistik*, various issues; and Fund staff estimates.
[1] National accounts basis. The general government comprises the Federal Government, the Länder governments, municipalities, the social security system, the Burden Equalization Fund, the European Recovery Program Fund, and the European Communities' Accounts.
[2] Estimated. Reflects the authorities' May 1988 tax estimates.
[3] Savings constitute the balance between current revenue and current expenditure; the financial balance is the difference between the sum of current expenditure plus capital transfers plus gross fixed investments and the sum of current revenue plus received capital transfers plus depreciation.
[4] Employment in the public sector (national accounts basis) in percent of total employment.

Table A25. Federal Republic of Germany: Territorial Authorities' Finances[1]

(Administrative basis)

	1987 Actual	1988 Estimate[2]	1985	1986	1987	1988 Estimate[2]
	(In billions of deutsche mark)		(Percentage change from preceding period)			
Total expenditures	**649.4**	**670.8**	**3.6**	**3.8**	**3.5**	**3.3**
Current expenditures	551.0	570.5	4.0	4.2	3.9	3.5
Wages and salaries	208.8	215.3	3.7	4.5	4.2	3.1
Goods	104.2	106.5	4.4	2.7	2.7	2.1
Interest	58.5	60.9	4.6	3.3	1.5	4.2
Current transfers	179.5	187.9	3.8	5.0	5.2	4.7
Capital expenditures	98.4	100.3	1.5	1.9	1.2	1.9
Investment	54.1	55.2	5.4	5.6	2.2	1.9
Capital transfers	23.3	23.1	−0.8	−0.8	−0.8	−0.9
Loans	21.0	22.1	−4.4	−3.9	0.9	5.1
Total revenue	**598.4**	**606.4**	**5.2**	**3.6**	**2.2**	**1.3**
Current revenue	577.6	586.5	5.5	3.3	2.1	1.5
Taxes	468.3	480.6	5.4	3.6	3.5	2.6
Other	109.4	105.9	6.0	2.4	−3.5	−3.1
Capital revenue	20.7	19.8	−3.1	11.2	6.5	−4.3
Financial balance	**−51.0**	**−64.5**	...	...	...	...
(In percent of GNP)	−2.5	−3.1	−2.1	−2.2	−2.5	−3.1

Sources: Data provided by the Ministry of Finance; and Fund staff estimates.

[1] Including the Federal Government, the Länder, the municipalities, the Burden Equalization Fund, the European Recovery Program Fund, and the European Communities' accounts.
[2] Reflects the authorities' May 1988 tax estimates.

Table A26. Federal Republic of Germany: Federal Government Budget
(Administrative basis)

	1987[1] Budget	1987[1] Actual	1988[2] Budget	1985	1986	1987	1988 Budget
	(In billions of deutsche mark)			(Percentage change from preceding period)			
Total expenditure	**268.5**	**269.0**	**275.4**	**2.1**	**1.7**	**2.9**	**2.4**
Current expenditure	234.1	234.4	241.7	2.7	2.5	3.2	3.1
Wages and salaries	39.2	39.3	40.0	3.4	3.5	3.6	1.8
Goods	40.9	40.1	40.5	1.2	2.5	1.5	1.1
Interest	30.9	31.0	32.3	5.1	3.8	2.5	4.1
Current transfers to other levels of government	23.0	23.9	24.4	3.1	3.1	3.6	2.0
Other current transfers	100.0	100.2	104.5	2.2	1.6	4.0	4.3
Capital expenditure	35.9	34.6	35.0	−1.3	−3.2	0.5	1.0
Investment	7.9	7.7	8.1	3.8	1.8	2.1	5.7
Capital transfers and loans to other levels of government	9.7	9.1	8.8	−0.1	−8.1	3.6	−3.2
Other capital transfers and loans	18.3	17.9	18.1	−3.8	−2.7	−1.7	1.2
Budget reserves	−1.4	...	−1.3	...	...	...	...
Total revenue	**241.4**	**240.7**	**235.7**	**5.0**	**1.7**	**1.2**	**−2.1**
Current revenue	235.5	235.6	230.2	5.3	1.1	0.7	−2.3
Taxes	216.0	217.0	217.4	4.6	1.3	3.9	0.2
Other[3]	19.5	18.6	12.8	11.5	−0.5	−26.1	−31.2
Capital revenue	5.9	5.1	5.5	−17.2	58.0	31.6	7.0
Financial balance	**−27.2**	**−28.4**	**−39.7**	...	...	...	...
(In percent of GNP)	−1.3	−1.4	−1.9	−1.3	−1.2	−1.4	−1.9

Sources: Data provided by the Ministry of Finance; and Fund staff estimates.
[1] Budgeted figures for all categories except taxes, which are estimates from May 1987.
[2] Budgeted figures for expenditures. Budgeted revenues amounted to DM 245.2 billion and the budgeted deficit to DM 29.9 billion. The revenues shown in the table reflect the authorities' May 1988 tax estimates.
[3] Excludes profits from the issue of coins.

Table A27. Federal Republic of Germany: Länder Government Finances

(Administrative basis)

	1987 Actual	1988 Estimate[1]	1985	1986	1987	1988 Estimate[1]
	(In billions of deutsche mark)		(Percentage change from preceding period)	Actual		
Total expenditures	263.0	270.2	3.9	4.0	3.9	2.7
Current expenditures	221.1	227.4	4.6	4.1	4.3	2.8
Wages and salaries	112.2	115.5	3.5	4.0	3.9	3.0
Goods	26.7	27.6	6.7	5.0	4.8	3.7
Interest	19.8	20.5	6.0	5.6	1.0	3.5
Current transfers to other levels of government	35.3	36.1	4.6	4.3	5.0	2.3
Other current transfers	27.2	27.6	5.8	2.6	7.3	1.5
Capital expenditures	41.9	42.8	0.4	3.4	1.7	2.1
Investment	11.2	11.3	6.1	2.6	3.5	0.6
Capital transfers to other levels of government	12.6	13.0	5.8	−10.1	7.2	3.5
Other capital transfers and loans	18.1	18.5	−6.7	14.8	−2.8	2.0
Total revenue	243.3	248.4	4.7	4.2	3.2	2.1
Current revenue	230.0	235.8	5.4	4.4	3.4	2.5
Taxes	173.2	177.9	6.0	4.9	3.8	2.7
Other current revenue	56.8	57.9	3.7	2.9	2.3	2.0
Capital revenue	13.3	12.7	−5.4	1.6	—	−5.0
Financial balance	−19.7	−21.7	...	...	...	...
(In percent of GNP)	−1.0	−1.0	−0.9	−0.9	−1.0	−1.0

Sources: Data provided by the Ministry of Finance; and Fund staff estimates.
[1] Reflects the authorities' May 1988 tax estimates.

Table A28. Federal Republic of Germany: Municipalities' Finances

(Administrative basis)

	1987 Actual	1988 Estimate[1]	1985	1986	1987	1988 Estimate[1]
	(In billions of deutsche mark)		(Percentage change from preceding period)	Actual		
Total expenditures	178.0	183.6	5.3	5.4	3.7	3.1
Current expenditures	136.6	141.5	5.3	5.6	4.0	3.6
Wages and salaries	57.4	59.2	4.3	6.1	5.3	3.2
Goods	37.5	38.5	6.3	1.4	2.5	2.7
Interest	7.5	7.6	−1.0	−3.5	−2.2	1.5
Current transfers to other levels of government	3.8	4.0	3.8	29.8	8.9	5.1
Other current transfers	30.4	32.2	8.1	10.6	4.2	5.8
Capital expenditures	41.4	42.0	5.4	4.7	2.7	1.6
Investment	35.2	35.7	5.5	7.5	1.7	1.4
Capital transfers and loans to other levels of government	1.5	1.6	5.3	−13.1	2.0	1.8
Other capital transfers and loans	4.6	4.8	4.7	−8.0	11.3	2.6
Total revenue	175.7	180.4	5.0	3.9	3.3	2.7
Current revenue	153.7	158.1	5.6	4.5	3.3	2.8
Taxes	59.8	61.6	7.1	4.6	1.7	3.0
Other current revenue	93.9	96.5	4.6	4.4	4.3	2.7
Capital revenue	21.9	22.3	1.3	0.3	3.5	1.8
Financial balance	−2.3	−3.2	...	...	...	...
(In percent of GNP)	−0.1	−0.2	—	−0.1	−0.1	−0.2

Sources: Data provided by the Ministry of Finance; and Fund staff estimates.
[1] Reflects the authorities' May 1988 tax estimates.

Table A29. Federal Republic of Germany: Territorial Authorities—Tax Revenue

(Cash basis)

	1970	1975	1980	1987	1987	1985	1986	1987
	(In percent)				(In billions of deutsche mark)	(Percentage change over preceding period)		
Total tax revenue	100.0	100.0	100.0	100.0	468.7	5.4	3.5	3.6
By type of tax								
Personal income tax	34.5	41.9	41.8	43.3	202.8	8.4	4.3	6.6
Corporation tax	5.7	4.2	5.8	5.8	27.3	21.0	1.5	−15.5
Wealth tax	1.9	1.4	1.3	1.2	5.4	−4.6	2.5	23.3
Trade tax[1]	7.0	7.4	7.4	6.7	31.4	8.6	4.0	−1.7
Value-added tax[2]	24.7	22.3	25.6	25.3	118.8	−0.6	1.2	6.9
Petroleum tax	7.5	7.1	5.8	5.6	26.1	2.0	4.6	1.9
Tobacco tax	4.2	3.7	3.1	3.1	14.5	0.2	0.2	0.2
Motor vehicle tax	2.5	2.2	1.8	1.8	8.4	0.9	27.3	−10.6
Other taxes	12.1	9.9	7.3	7.2	33.9	2.4	3.6	3.2
By level of government								
Federal Government[3]	55.2	50.1	48.7	46.7	218.8	4.6	1.3	3.9
Länder[4]	32.7	33.7	34.4	35.5	166.5	6.0	5.0	4.0
Municipalities[5]	11.8	13.7	14.0	13.8	64.9	7.6	3.8	3.1
European Communities	—	2.5	2.9	3.9	18.3	3.1	18.2	2.1
Memorandum item								
Tax revenue (*in percent of GNP*)	22.8	23.5	24.6	23.2		23.7	23.2	23.2

Sources: Deutsche Bundesbank, *Monthly Report,* various issues; and data provided by Ministry of Finance.
[1] Tax based on capital stock and on return to capital.
[2] Including turnover tax on importers.
[3] Including Burden Equalization Fund.
[4] Excluding municipal taxes of the Länder Berlin, Bremen, and Hamburg.
[5] Including municipal taxes of Berlin, Bremen, and Hamburg.

Table A30. Federal Republic of Germany: General Government—Revenue and Expenditure
(National accounts basis)

	1987 Actual	1988 Estimate[1]	1985	1986	1987	1988 Estimate[1]
	(In billions of deutsche mark)		(Percentage change from preceding period)			
Total expenditure	**946.2**	**985.3**	**3.2**	**3.9**	**3.9**	**4.1**
Expenditures on goods and services	444.4	460.2	4.1	5.0	3.7	3.6
Public consumption	396.8	410.9	4.4	4.5	3.8	3.6
Public investment	47.6	49.3	2.2	9.1	2.4	3.6
Transfer payments	501.8	525.1	2.3	3.0	4.2	4.6
Social benefits	327.4	342.0	2.2	3.7	4.8	4.5
Subsidies	43.6	44.5	3.9	9.5	5.8	2.1
Interest	57.9	60.2	4.9	3.5	1.2	4.0
Other	72.9	78.4	0.2	−3.4	2.8	7.5
Total revenue	**912.0**	**936.5**	**4.9**	**3.7**	**2.8**	**2.7**
Tax revenue	491.7	503.7	4.7	2.9	3.9	2.4
Direct taxes	246.2	253.0	7.8	3.2	3.5	2.1
Indirect taxes	245.5	250.7	1.8	2.6	2.6	2.8
Social security contributions	350.6	367.5	5.0	5.5	3.9	4.8
Other revenue	69.7	65.3	4.9	1.4	−7.8	−6.3
Financial balance	**−34.2**	**−48.8**	...	...	...	...
Of which:						
Territorial authorities	−41.1	−52.8	...	...	...	...
Social security system	6.9	4.0	...	...	...	...
Memorandum items (in percent of GNP)						
Financial balance of general government	−1.7	−2.3	−1.1	−1.2	−1.7	−2.3
Of which:						
Territorial authorities	−2.0	−2.5	−1.4	−1.7	−2.0	−2.5
Social security system	0.3	0.2	0.3	0.5	0.3	0.2

Sources: Data provided by the Ministry of Finance; and Fund staff estimates.
[1] Reflects the authorities' May 1988 tax estimates.

Table A31. Federal Republic of Germany: Outstanding Debt of the Territorial Authorities

(In billions of deutsche mark)

	1987	1987 (In percent)	1985	1986	1987
			Changes during period		
By borrower					
Federal Government[1]	440.5	51.9	25.1	23.0	25.1
ERP Fund[2]	5.9	0.7	0.1	−0.3	−0.5
Länder	284.4	33.5	16.9	16.9	20.0
Municipalities	117.6	13.9	0.6	1.1	2.8
Total	848.4	100.0	42.7	40.8	47.4
By type of instrument					
Bundesbank credit	0.8	0.1	−2.3	2.8	−2.1
Non-interest-bearing treasury bills	5.5	0.6	−0.3	−1.7	−2.7
Medium-term notes	46.9	5.5	3.8	7.6	13.7
Federal bonds[3]	84.4	10.0	10.7	8.2	4.3
Federal savings bonds[4]	31.1	3.7	4.8	2.2	3.0
Government bonds	171.7	20.2	19.3	33.5	28.3
Bank credits	461.5	54.4	8.7	−8.7	5.3
Social security system loans	7.6	0.9	−0.7	−0.9	−0.6
Other[5]	38.7	4.6	−1.3	−2.2	−1.7
Total	848.4	100.0	42.7	40.8	47.4
By type of creditor					
Bundesbank	12.6	1.5	−2.4	3.7	−2.7
Banks	493.7	58.2	20.2	−0.5	26.6
Social security system	7.6	0.9	−0.7	−0.9	−0.6
Foreign lenders	176.8	20.8	17.6	36.6	17.9
Other	157.6	18.6	8.0	1.9	6.2
Total	848.4	100.0	42.7	40.8	47.4
(In percent of GNP)	41.9		0.7	−0.1	0.8
Memorandum items					
Debt of the Federal Railways	40.8		0.4	1.9	2.8
Of which: Foreign debt	*13.0*		*1.5*	*3.1*	*1.9*
Debt of the Federal Postal System	61.1		4.8	4.7	6.1
Of which: Foreign debt	*10.4*		*1.9*	*4.4*	*1.2*

Source: Deutsche Bundesbank, *Monthly Report,* various issues.

[1] Including Burden Equalization Fund.
[2] European Recovery Program Fund.
[3] Federal bonds (*Bundesobligationen*) are fixed interest, five-year securities available in small denominations to the general public.
[4] Federal savings bonds (*Bundesschatzbriefe*) are six- or seven-year securities with a rising interest rate schedule. They are redeemable at face value before maturity.
[5] Including loans raised abroad.

Table A32. General Government Current Receipts of Major Industrial Countries

(In percent of GDP)

	1960	1968	1974	1975	1976	1977	1978	1979	1980	1981	1982	1983	1984	1985
United States	26.3	28.7	30.3	28.8	29.5	29.7	29.9	30.5	30.8	31.6	31.1	30.7	30.7	31.1
Japan	20.7	19.6	24.5	24.0	23.6	24.7	24.5	26.3	27.6	29.1	29.5	29.8	30.4	31.2
Germany, Federal Republic of	35.0	37.8	42.7	42.7	44.0	45.0	44.7	44.4	44.7	44.8	45.4	45.1	45.4	45.4
France	34.9	38.8	39.4	40.3	42.5	42.4	42.3	43.7	45.5	46.2	47.1	47.7	48.5	48.5
United Kingdom	30.0	37.6	39.7	40.2	39.3	38.9	37.5	38.3	40.1	42.2	43.3	42.3	42.8	...
Italy	28.8	31.6	30.6	31.2	32.9	34.3	36.0	35.7	37.8	39.3	42.0	45.0	44.2	44.1
Canada	26.0	31.7	37.2	36.1	35.8	36.1	35.7	35.5	36.2	38.5	39.0	38.7	38.9	38.9
Total	27.8	30.3	32.7	32.2	32.7	33.0	32.9	33.9	34.9	35.3	35.4	35.1	34.9	34.7
Total EC	31.5	35.9	38.4	39.0	40.2	40.9	41.0	41.3	42.3	43.2	44.3	44.9	45.2	46.9
Total OECD-Europe	31.3	35.8	38.7	39.5	40.9	41.6	41.6	41.7	42.8	43.6	44.5	45.1	45.4	47.0
Total OECD less United States	29.5	32.4	35.2	35.8	36.5	37.0	36.5	37.5	38.8	39.2	40.1	40.1	40.2	40.8
Total OECD	27.7	30.6	33.3	33.1	33.7	34.1	34.1	35.0	36.0	36.2	36.4	36.1	35.9	35.9

Source: Organization for Economic Cooperation and Development, *Historical Statistics 1960–1986* (Paris, 1987).

APPENDIX I • STATISTICAL TABLES

Table A33. Federal Republic of Germany: The 1990 Tax Cuts[1]
(In billions of deutsche mark)

	On Accrual Basis								On Cash Basis								
	1990				1990				1991				1992				
	Total	FG	SG	M	Total	FG	SG	M	Total	FG	SG	M	Total	FG	SG	M	
Tax rate reform																	
Increase of basic allowance from DM 4,752/9,504 to DM 5,616/11,232 (unmarried/married)	5.9	2.5	2.6	0.8	5.4	2.3	2.4	0.8	5.9	2.5	2.6	0.8	6.0	2.6	2.6	0.8	
Reduction of tax entry rate from 22 to 19 percent	6.2	2.6	2.7	0.9	5.7	2.4	2.5	0.8	6.2	2.6	2.7	0.9	6.3	2.7	2.7	0.9	
Flattening of tax rate schedule	18.7	7.9	8.2	2.6	16.4	7.0	7.2	2.3	19.3	8.2	8.4	2.7	20.9	8.9	9.1	2.9	
Reduction of the top marginal income tax rate from 56 to 53 percent	1.1	0.5	0.5	0.2	0.9	0.4	0.4	0.1	1.2	0.5	0.5	0.2	1.3	0.6	0.6	0.2	
Relief for families																	
Increase in child allowance from DM 2,484 to DM 3,042	1.8	0.8	0.8	0.3	1.7	0.7	0.7	0.3	1.8	0.8	0.8	0.3	1.8	0.8	0.8	0.3	
Other family deductions	0.4	0.2	0.2	0.1	0.3	0.1	0.1	—	0.4	0.2	0.2	0.1	0.4	0.2	0.2	0.1	
Increase in tax deductions for provisionary expenses	0.6	0.3	0.3	0.1	0.3	0.1	0.1	—	0.5	0.2	0.2	0.1	0.6	0.3	0.3	0.1	
Reduction of corporate taxation																	
For retained earnings from 56 to 50 percent	2.1	1.1	1.1	—	1.9	1.0	1.0	—	2.2	1.1	1.1	—	2.2	1.1	1.1	—	
For corporations subject to a reduced rate from 50 to 46 percent	0.4	0.2	0.2	—	0.4	0.2	0.2	—	0.4	0.2	0.2	—	0.4	0.2	0.2	—	
Total	37.2	16.0	16.4	4.8	32.9	14.2	14.5	4.3	37.8	16.3	16.7	4.9	39.9	17.1	17.6	5.2	
Increase in tax revenue owing to cuts in tax exemptions[2]	18.1	7.6	8.8	2.8	13.4	6.0	5.8	1.6	13.9	6.2	6.1	1.7	16.4	7.2	7.1	2.1	
"Net" tax cuts	19.1	8.3	8.8	2.0	19.5	8.2	8.7	2.7	23.9	10.1	10.6	3.2	23.5	10.0	10.5	3.0	

Source: Presse und Informationsamt der Bundesregierung, "Steuerreform 1986, 1988, 1990." *Aktuelle Beiträge zur Wirtschafts und Finanzpolitik*, Nr. 45/1988 (Bonn, July 8, 1988).

[1] Totals may differ from sums of components because of rounding; FG = Federal Government; SG = state governments (including city states); M = municipalities.
[2] See Table A34 for breakdown of these figures.

Table A34. Federal Republic of Germany: The 1990 Cuts in Tax Exemptions[1]

(In billions of deutsche mark)

	On Accrual Basis												On Cash Basis												
	1990				1990				1991				1991				1992								
	Total	FG	SG	M	Total	FG	SG	M	Total	FG	SG	M	Total	FG	SG	M	Total	FG	SG	M					
Income tax	9.7	4.4	4.4	0.9	7.7	3.6	3.6	0.6	9.1	4.1	4.2	0.7	9.8	4.4	4.5	0.9									
Of which:																									
Substitution of specific tax deductions for employees by a general allowance	1.2	0.5	0.5	0.2	1.0	0.4	0.4	0.1	1.2	0.5	0.5	0.2	1.2	0.5	0.5	0.2									
Introduction of a 10 percent withholding tax on interest income	4.2	2.1	2.1	—	4.0	2.0	2.0	—	4.3	2.2	2.2	—	4.8	2.4	2.4	—									
Corporation tax	0.1	—	0.1	—	—	—	—	—	—	—	—	—	0.1	—	0.1	—									
Trade tax	0.3	−0.1	−0.1	0.4	0.2	−0.1	—	0.3	0.2	−0.1	—	0.3	0.3	−0.1	−0.1	0.4									
Other taxes	2.5	0.9	0.8	0.8	0.9	0.5	0.3	0.1	1.0	0.5	0.3	0.2	1.3	0.6	0.4	0.3									
Of which:																									
Interest on uncollected taxes and tax advances	0.9	0.3	0.3	0.3	—	—	—	—	—	—	—	—	—	—	—	—									
Special provisions	2.5	1.2	1.2	0.1	0.3	0.2	0.1	—	1.6	0.8	0.7	0.1	2.5	1.2	1.2	0.1									
Of which:																									
Abrogation of the investment incentives law	1.6	0.8	0.8	—	0.2	0.1	0.1	—	1.1	0.5	0.5	—	1.6	0.8	0.8	—									
Legal or administrative ordinances	2.1	0.9	0.9	0.3	1.8	0.8	0.8	0.3	2.0	0.8	0.8	0.3	2.1	0.9	0.9	0.3									
Of which:																									
Abrogation of tax deductions for subsidized company meals	1.0	0.5	0.4	0.1	0.9	0.4	0.4	0.1	1.0	0.4	0.4	0.1	1.0	0.4	0.4	0.1									
Measures that expire and are not renewed[2]	1.0	0.4	0.4	0.2	...	...	...	...	...	...	...	...	0.3	0.1	0.1	0.1									
Total	18.1	7.6	7.6	2.8	11.1	5.0	4.8	1.3	13.9	6.2	6.1	1.7	16.4	7.2	7.1	2.1									

Source: Presse und Informationsamt der Bundesregierung, "Steuerreform 1986, 1988, 1990." *Aktuelle Beiträge zur Wirtschafts und Finanzpolitik,* Nr. 45/1988 (Bonn, July 8, 1988).

[1] Totals may differ from sums of components because of rounding; FG = Federal Government; SG = state governments (including city states); M = municipalities.
[2] Effective from 1991.

Table A35. International Comparison of Recent and Proposed Changes in Personal Income Tax Systems

	Including Social Security Marginal tax rates on average wages under present tax systems[1]	Excluding Social Security Top marginal rate[2]		
		Previous	Present	Proposed
Australia	47.3	60.0	55.0	49.0
Austria	54.5	...	62.0	50.0
Belgium	62.7	...	86.7	...
Canada	33.7	63.6	51.3	29.0
Denmark	62.4	73.0	68.0	...
Finland	53.2	...	68.5	...
France	51.2	65.0	58.0	50.0
Germany, Federal Republic of	62.7	...	56.0	53.0
Greece	40.1	60.0	63.0	...
Iceland	n.a.	...	55.6	...
Ireland	61.3	65.0	58.0	...
Italy	57.8	76.0	65.0	56.0
Japan	31.5	88.0	76.5	50.0
Luxembourg	53.6	57.0	56.0	...
Netherlands	61.9	60.0	72.0	...
New Zealand	30.0	66.0	48.0	...
Norway	60.1	71.0	57.6	...
Portugal	35.9	84.4	68.8	...
Spain	52.8	68.5	66.0	...
Sweden	62.0	87.7	77.4	...
Switzerland	39.4	...	45.8	...
Turkey	n.a.	78.0	50.0	...
United Kingdom	43.9	83.0	60.0	40.0
United States	40.9	75.0	38.0	33.0
Average	50.0	71.8	60.8	45.6

Sources: Organization for Economic Cooperation and Development, *Economic Outlook* (Paris), Vol. 41 (June 1987), p. 22; and Vito Tanzi, "Tax Reform in Industrial Countries and the Impact of the U.S. Tax Reform Act of 1986," IMF Working Paper, No. 87/61 (Washington: International Monetary Fund, 1987).

[1] Overall marginal tax rate for an average (unmarried) production worker, allowing for direct taxes at all levels of government, social security contributions by both employers and employees, and relevant tax concessions. The major data source is OECD, *The Tax/Benefit Position of Production Workers 1981–1985* (Paris, 1986). The figures shown are estimates for 1986.

[2] Global effective rate (excluding social security contributions) but allowing for deductibility of taxes paid to lower levels of government.

Table A36. International Comparison of Social Security Transfers as a Percentage of GDP
(In percent of GDP)

	Average			1980	1981	1982	1983	1984	1985
	1968–73	1974–79	1980–85						
United States	7.7	10.3	11.3	10.9	11.1	11.9	11.9	11.0	11.0
Japan	4.8	8.4	10.8	10.1	10.6	11.0	11.3	11.1	11.0
Germany, Federal Republic of	13.2	16.7	16.8	16.5	17.2	17.6	17.0	16.5	16.1
France	17.2	21.0	25.4	23.2	24.6	25.7	26.0	26.3	26.4
United Kingdom	8.8	10.5	13.3	11.5	12.9	13.9	13.8	14.0	13.8
Italy	13.0	15.4	18.5	15.8	17.7	18.7	19.9	19.4	19.5
Canada	8.2	9.9	11.4	9.9	9.9	11.8	12.4	12.2	12.3
Total	9.0	12.1	13.4	13.1	13.2	14.0	13.9	13.2	13.0
Austria	15.6	17.8	19.8	19.0	19.5	19.9	20.1	20.0	20.2
Belgium	14.4	19.2	22.2	20.9	22.6	22.4	22.9	22.5	22.0
Denmark	11.1	14.0	17.3	16.6	17.8	18.1	17.8	17.1	16.3
Finland	7.4	8.9	9.7	8.7	8.8	9.7	10.2	10.2	10.7
Greece	7.8	8.1	12.7	9.2	11.0	13.0	13.4	14.0	15.0
Iceland	9.4	9.8	4.8	4.5	4.8	5.0	4.6	4.8	5.0
Ireland	8.6	11.9	...	12.6	13.6	15.4	16.2	16.2	...
Luxembourg	14.7	20.0	...	22.7	23.8	23.2	22.3	21.5	...
Netherlands	17.7	23.5	27.5	25.9	26.9	28.4	28.8	27.7	27.0
Norway	12.4	14.2	14.9	14.4	14.5	15.0	15.5	15.0	14.8
Portugal	3.5	9.3	...	10.6	11.6	...	...	...	...
Spain	8.9	11.7	15.7	14.2	15.6	15.5	16.1	15.9	16.7
Sweden	11.6	15.9	18.2	17.8	18.4	18.6	18.5	17.6	18.2
Switzerland	8.5	12.6	13.3	12.7	12.4	13.2	13.5	14.1	13.7
Turkey	1.5	2.8	3.4	3.0	3.2	3.4	3.9	3.5	3.4
Smaller European	11.2	15.5	18.1	17.0	17.7	18.3	18.7	18.2	18.6
Australia	5.6	8.3	7.7	8.3	8.5	9.6	9.9	10.0	9.7
New Zealand	...	...	...	...	...	...	...	...	...
Total	10.4	14.5	16.9	15.9	16.2	16.9	17.2	16.7	18.6
Total EC	13.0	16.4	19.1	17.4	18.5	19.3	19.4	19.3	20.7
Total OECD – Europe	12.7	16.2	18.6	17.1	18.1	18.8	19.0	18.8	19.9
Total OECD less United States	10.6	13.8	15.7	14.9	15.3	16.1	16.1	15.8	16.1
Total OECD	9.2	12.4	13.8	13.5	13.7	14.4	14.3	13.6	13.5

Source: Organization for Economic Cooperation and Development, *Historical Statistics 1960–1986* (Paris, 1987).

Table A37. Dependency Ratios in Seven Major Countries[1]

	1950	1960	1970	1980	1990	2000	2010	2020
Canada	12.2	12.7	12.7	13.1	15.8	17.3	18.7	25.0
France	17.3	18.8	20.7	21.4	19.6	22.4	22.2	28.3
Germany, Federal Republic of	13.0	16.0	20.7	22.7	21.1	24.4	31.1	33.4
Italy	12.6	14.2	16.4	20.9	21.4	24.7	26.3	29.2
Japan	8.3	9.0	10.3	13.4	15.9	22.0	28.2	34.0
United Kingdom	16.0	17.9	20.6	23.1	22.9	22.6	23.0	26.8
United States	12.5	15.5	15.8	17.1	17.9	17.6	18.0	23.9
Average	13.3	14.9	16.7	18.8	19.2	21.6	23.9	28.7

Source: Organization for Economic Cooperation and Development, *Structural Adjustment and Economic Performance* (Paris, 1987), p. 331.

[1] Ratio of population aged 65 and over to population aged 15–64.

Table A38. Federal Republic of Germany: Expenditures of the Medical Insurance System and Development of Macroeconomic Aggregates

	Expenditures of Medical Insurance	Total Social Transfers	Nominal GNP	Gross Wage and Salary Income
(1970 = 100)				
1975	242	192	152	158
1980	357	264	220	223
1986	475	336	288	272
(In annual percentage changes)				
1971–75	19.3	13.9	8.7	9.6
1976–80	8.1	6.6	7.7	7.1
1981–86	4.9	4.1	4.6	3.4

Source: A. Seffen, "Entwicklung der Krankenversicherungsfinanzen seit 1970," *IW-Trends*, No. 4 (December 20, 1987).

Table A39. Federal Republic of Germany: Revenues and Expenditures of the Medical Insurance System

	Revenues	Expenditures	Balance	Rate of Contribution[1]
	(In billions of deutsche mark)			*(In percent)*
1970	26.12	25.18	0.94	8.20
1971	31.28	31.14	0.14	8.13
1972	36.21	36.40	−0.19	8.39
1973	44.46	43.37	1.09	9.15
1974	51.11	51.81	−0.70	9.47
1975	60.74	60.99	−0.25	10.43
1976	70.20	66.56	3.64	11.28
1977	73.50	69.82	3.68	11.37
1978	76.43	74.79	1.64	11.41
1979	80.83	81.06	−0.23	11.26
1980	88.45	89.83	−1.38	11.38
1981	96.49	96.39	0.10	11.79
1982	101.71	97.22	4.49	12.00
1983	103.48	100.69	2.79	11.83
1984	105.81	108.68	−2.87	11.44
1985	111.83	114.11	−2.28	11.80
1986	118.29	119.58	−1.29	12.19
1987[2]	124.50	124.40	0.10	12.60

Source: A. Seffen, "Entwicklung der Krankenversicherungsfinanzen seit 1970," *IW-Trends*, No. 4 (December 20, 1987).

[1] On July 1; sum of contributions of employers and employees.
[2] Provisional.

Table A40. Federal Republic of Germany: Nominal and Effective Protection in 1982[1]
(In percent)

Product Group	Nominal Protection		Effective Protection		
	Tariff protection	Explicit and implicit protection[2]	Explicit and implicit protection[2]	Degree of subsidization	Total effective protection
Coal mining	—	44.2	189.2	147.6	336.8
Basic materials					
Chemicals	9.9	9.9	14.3	5.0	19.3
Mineral oil products	4.0	4.0	14.0	0.8	14.8
Stone, sand, and gravel	4.9	4.9	5.4	1.4	6.8
Iron and steel	6.4	20.0	43.1	14.9	58.0
Nonferrous metals	7.0	7.0	13.5	4.2	17.7
Iron and steel castings	6.9	6.9	7.1	0.8	7.9
Hoops and strips of iron and steel	7.5	7.5	1.9	0.6	2.5
Wood	5.7	5.7	16.7	2.5	19.2
Paper, paperboard, and pulp	10.7	10.7	29.2	2.1	31.3
Rubber products	7.1	7.1	6.6	0.7	7.3
Investment goods					
Structures of steel and metal	5.0	5.0	−0.7	1.9	1.2
Machinery	5.6	5.6	1.7	2.9	4.6
Office equipment	5.9	5.9	7.4	4.6	12.0
Road vehicles	10.3	10.3	9.9	1.0	10.9
Ships and boats	2.7	2.7	−6.5	26.0	19.5
Aircraft and spacecraft	7.2	7.2	15.8	29.6	45.4
Electronics	7.0	7.0	6.0	3.6	9.6
Iron, sheet metal, and metal products	6.8	6.8	3.7	1.6	5.3
Consumer goods					
Plastic products	12.0	12.0	17.8	2.1	19.9
Ceramic products	10.3	10.3	13.0	2.4	15.4
Glass and glassware	8.8	8.8	12.0	1.5	13.5
Mechanical and optical products, watches	8.0	8.0	7.6	2.0	9.6
Musical instruments, toys, sports equipment, and jewelry	8.2	8.2	8.6	1.4	10.0
Wood products	7.1	7.1	7.5	1.2	8.7
Paper and paperboard products	12.7	12.7	26.0	4.6	30.6
Printing and publishing	3.8	3.8	0.7	3.6	4.3
Leather and leather products	7.9	7.9	8.5	0.7	9.2
Textiles	13.0	34.4	71.2	2.1	73.3
Clothing	15.3	44.7	120.0	2.9	122.9
Average	7.9	11.2	22.4	9.2	31.6
Standard deviation	2.9	10.8	39.8	27.0	62.9
Coefficient of variation	0.4	1.0	1.8	2.9	2.0

Source: D. Witteler, "Tarifäre und nichttarifäre Handelshemmnisse in der Bundesrepublik Deutschland—Ausmass und Ursachen," *Die Weltwirtschaft* (Tübingen), No. 1 (1986), p. 140.

[1] Against countries outside the EC and the European Tree Trade Association.
[2] Includes the tariff equivalents of nontariff barriers.

Table A41. Federal Republic of Germany: Nontariff Barriers to Trade in 1982

(Share of affected product lines in percent of all product lines of the foreign trade statistics)

Product	Import Restrictions[1]	Countervailing Duties and Voluntary Price Increases	Minimum Pricing
Manufacturing and mining	11.7	1.4	1.1
Coal mining	45.5	1.7	—
Plastic products	—	3.0	—
Rubber products	4.6	—	—
Stone, sand, and gravel	2.7	—	—
Ceramic products[2]	28.0	—	—
Glass and glassware	9.1	—	—
Iron and steel	13.8	16.4	26.6
Nonferrous metals	3.0	—	—
Iron and steel castings	2.9	—	—
Hoops and strips of iron or steel	—	1.6	—
Machinery	1.0	0.1	—
Aircraft and spacecraft	2.3	—	—
Electronics	0.9	1.3	—
Mechanical and optical products, watches	0.4	2.3	—
Iron, sheet metal, and metal products	0.6	—	—
Musical instruments, toys, sports equipment, and jewelry	3.3	—	—
Wood	20.5	5.1	—
Wood products	5.9	5.9	—
Leather and leather products	39.4	—	—
Textiles	65.0	—	—
Clothing	56.5	—	—

Source: D. Witteler, "Tarifäre und nichttarifäre Handelshemmnisse in der Bundesrepublik Deutschland—Ausmass und Ursachen," *Die Weltwirtschaft* (Tübingen), No. 2 (December 1987), p. 142.

[1] Unilateral restrictions and voluntary export restraints for textiles and clothing.
[2] The restriction of imports from Japan was lifted in 1985.

Table A42. Import Restrictions in the EC in 1982–83 Under Article 115 of the EEC Treaty[1]

(Number of affected products)[2]

	Textiles and clothing[3]	Others[4]	Total
Belgium/Luxembourg	162	70	232
Denmark	56	51	107
France	208	340	548
Germany, Federal Republic of	36	11	47
Greece	6	247	253
Ireland	64	98	162
Italy	107	459	566
United Kingdom	278	8	286

Source: D. Witteler, "Tarifäre und nichttarifäre Handelshemmnisse in der Bundesrepublik Deutschland—Ausmass und Ursachen," *Die Weltwirtschaft* (Tübingen), No. 1 (1986), p. 142.

[1] Article 115 of the EEC Treaty allows member countries to impose national restrictions against imports from third countries under certain circumstances.
[2] Product lines of trade statistics.
[3] Against Multifiber Arrangement countries.
[4] Excluding state-trading countries.

Table A43. International Comparison of Subsidies[1]

(In percent of GDP)

	1975	1977	1979	1980	1981	1982	1983	1984	1985	1986[2]
Belgium	1.2	1.4	1.7	1.4	1.5	1.4	1.4	1.5	1.5	1.4
Germany, Federal Republic of	2.0	2.1	2.2	2.1	1.9	1.8	1.9	2.1	2.0	2.1
Denmark	2.8	3.2	3.2	3.2	3.0	3.2	3.3	3.4	3.3	3.0
France	2.0	2.2	2.0	1.9	2.2	2.2	2.1	2.4	2.3	2.9
United Kingdom	3.5	2.2	2.3	2.5	2.6	2.1	2.1	2.4	2.2	1.7
Italy	2.2	2.3	2.6	2.4	2.5	3.0	2.5	2.7	2.7	3.0
Japan	1.5	1.3	1.3	1.5	1.5	1.4	1.4	1.3	1.2	1.1
Canada	2.5	1.8	2.0	2.7	2.7	2.5	2.5	2.8	2.5	2.0
Netherlands	1.0	1.4	1.3	1.5	1.6	1.7	1.8	1.9	2.0	3.1
Norway	6.3	7.4	7.0	7.0	6.7	6.5	6.1	5.7	5.4	5.8
Austria	2.9	2.9	2.9	3.0	3.0	3.0	3.1	3.0	2.7	3.0
Sweden	3.1	4.1	4.3	4.3	4.7	5.0	5.2	5.0	4.9	4.8
Switzerland	1.2	1.4	1.4	1.4	1.2	1.3	1.4	1.4	1.4	1.4
United States	0.3	0.4	0.4	0.4	0.4	0.5	0.7	0.6	0.6	0.6

Source: Ministry of Finance, *Eleventh Subsidy Report* (Bonn, November 1987).

[1] Financial assistance of general government.
[2] Partly estimated.

Table A44. Federal Republic of Germany: Subsidies

(In billions of deutsche mark)

	1970	1975	1980	1984	1985	1986	1987	1988 Budget
Federal Government								
Payments	8.1	10.9	13.5	13.6	13.4[1]	14.3	14.6	15.1
Preferential tax treatment	6.2	9.7	12.1	15.1	15.7	15.7	16.0	16.7
Total	14.3	20.6	25.6	28.7	29.1	30.0	30.6	31.8
(In percent of GNP)	2.1	2.0	1.7	1.6	1.6	1.5	1.5	1.5
Länder and municipalities								
Payments[2]	7.1	9.1	14.0	14.5	14.0[1]	13.3	14.2	. . .
Preferential tax treatment	6.9	12.2	15.1	19.2	18.3	18.4	18.5	19.5
Total	14.0	21.3	29.1	33.7	32.3	31.7	32.7	. . .
(In percent of GNP)	2.1	2.1	2.0	1.9	1.8	1.6	1.6	. . .
ERP payments[3]	1.1	1.3	2.7	3.0	2.9	3.2	3.2	. . .
EC payments	2.9	2.2	6.2	7.5	8.0	9.4	8.4	. . .
Total subsidies	32.3	45.4	63.6	72.9	72.3	74.3	74.9	. . .
(In percent of GNP)	4.8	4.4	4.3	4.1	3.9	3.8	3.7	. . .

Source: Ministry of Finance, *Eleventh Subsidy Report* (Bonn, November 1987).
[1] In 1985, DM 0.8 billion in subsidies for housing was permanently shifted from the Länder to the Federal Government.
[2] Estimates.
[3] Lending mainly to small and medium-sized enterprises by the European Recovery Program.

APPENDIX I • STATISTICAL TABLES

Table A45. Federal Republic of Germany: Fiscal Assistance and Tax Relief of the Federal Government
(In billions of deutsche mark)

	1970	1975	1980	1981	1982	1983	1984	1985	1986	1987	1988[1]
Agriculture and forestry	4.8	4.2	3.7	2.8	2.7	2.6	3.8[2]	4.7[2]	4.9[2]	5.2	5.7
(In percent of value added)	21.8	14.7	12.1	8.7	7.3	8.1	11.0	14.9	14.6	17.0	...
Enterprises (excluding transport)	3.7	5.6	9.2	9.1	8.9	9.6	11.4	11.1	11.0	11.7	13.2
(In percent of value added)	1.3	1.4	1.7	1.7	1.6	1.6	1.9	1.7	1.6	1.6	...
Of which:											
Mining	*0.5*	*1.0*	*2.6*	*2.2*	*1.4*	*1.4*	*2.0*	*1.6*	*2.0*	*3.0*	*3.3*
Energy and raw materials	—	*0.4*	*0.3*	*0.4*	*0.5*	*0.4*	*0.4*	*0.4*	*0.4*	*0.3*	*0.3*
Technology and innovation	*0.2*	*0.1*	*0.5*	*0.6*	*0.7*	*0.8*	*0.9*	*0.9*	*1.0*	*0.8*	*0.8*
Sectoral aid	*0.2*	*0.4*	*0.7*	*0.8*	*1.0*	*0.8*	*1.6*	*1.3*	*0.6*	*0.5*	*1.2*
Shipbuilding	—	*0.1*	*0.3*	*0.3*	*0.2*	*0.3*	*0.2*	*0.1*	*0.1*	*0.1*	*0.3*
Aircraft	*0.2*	*0.3*	*0.4*	*0.4*	*0.4*	*0.2*	*0.3*	*0.4*	*0.4*	*0.4*	*0.9*
Steel	—	—	—	*0.1*	*0.4*	*0.3*	*1.1*	*0.7*	*0.1*	—	—
Regional measures	*2.1*	*3.0*	*4.2*	*4.4*	*4.5*	*4.8*	*4.8*	*5.4*	*5.7*	*5.9*	*6.0*
Banking	*0.2*	*0.3*	*0.2*	—	—	—	—	—	—	—	—
Other measures	*0.6*	*0.4*	*0.6*	*0.8*	*0.8*	*1.4*	*1.6*	*1.5*	*1.4*	*1.3*	*1.6*
Transport	0.9	1.4	2.5	2.2	2.0	1.9	1.8	1.7	1.7	1.7	1.7
(In percent of value added)	2.4	2.3	2.9	2.4	2.1	2.0	1.8	1.6	1.6	1.6	...
Construction[3]	1.6	3.5	4.8	5.7	5.9	6.1	6.8	6.8	7.3	7.5	7.2
(In percent of value added)	3.1	5.5	4.8	5.8	6.2	6.3	6.9	7.3	7.4	7.4	...
Savings and capital formation	2.7	5.1	4.0	4.0	4.0	4.2	3.2	3.2	3.0	2.7	2.2
(In percent of business sector value added)	0.5	0.6	0.3	0.3	0.3	0.3	0.2	0.2	0.2	0.2	...
Other	0.6	0.9	1.5	1.8	1.7	1.6	1.6	1.6	2.0	1.8	1.8
Total	14.3	20.6	25.5	25.7	25.1	26.0	28.7	29.1	30.0	30.6	31.8
(In percent of business sector value added)	2.5	2.4	2.1	2.0	1.9	1.9	1.9	1.9	1.8	1.8	...
Memorandum item											
EC payments for agricultural price support	2.9	2.2	6.2	5.3	4.9	7.0	7.5	8.0	9.4	8.4	9.9
(In percent of value added)	13.1	7.6	20.5	16.9	13.5	21.8	21.5	25.2	27.8	27.4	...

Source: Ministry of Finance, *Eleventh Subsidy Report* (Bonn, November 1987).
[1] Budget.
[2] Including the effect of changes in the value-added tax for inputs in agriculture.
[3] In 1985, DM 0.8 billion in subsidies for housing was permanently shifted from the Länder to the Federal Government.

Table A46. Federal Republic of Germany: Subsidies Provided by the Federal Government for Selected Industries
(In deutsche mark per employee)

	1966	1970	1975	1980	1981	1982	1983	1984	1985
Agriculture, forestry, and fishing	1,182	2,102	2,353	2,556	1,954	1,905	1,876	2,786	3,486
Mining	915	1,716	3,934	11,056	9,055	6,097	5,887	9,236	7,460
Iron and steel	8	13	14	159	277	1,425	1,169	4,636	2,991
Shipbuilding	—	368	1,351	4,379	5,190	4,000	5,273	3,553	3,289
Air and space industry	765	3,659	6,341	6,764	7,517	6,948	4,246	5,518	7,105
Shipping[1]	860	994	1,072	1,290	1,205	1,007	1,020	1,010	884
Total	329	537	802	971	983	979	1,028	1,134	1,142

Source: Ministry of Finance, *Eleventh Subsidy Report* (Bonn, November 1987).
[1] Including waterways, ports, and other sea traffic.

Table A47. Key Features of the Agricultural Sector in the EC in 1986

	France	Federal Republic of Germany	Italy	United Kingdom	EC
Utilized agricultural area (*1000 hectares*)	31,418	12,000	17,445	18,612	97,279
Land per holding (1985; *in hectares*)	27.0	16.0	5.6	65.1	13.9
Number of holdings (1985; *in 1000s*)	1,057	740	2,801	258	6,359
Employment (*in percent of employed civilian work force*)	7.3	5.3	10.9	2.6	7.0
Value added (1985; *in percent of GDP*)	3.7	1.8	5.0	1.8	3.4
Imports (*in percent of total imports*)	15.8	13.6	16.2	14.8	15.3
Exports (*in percent of total exports*)	12.6	3.6	5.2	6.7	8.2
Agricultural trade balance (*in millions of ECU*)	−824	−7,707	−5,015	−5,560	−22,676

Source: Commission of the European Communities, *The Agricultural Situation in the Community—1987 Report* (Brussels; Luxembourg, 1988).

Table A48. Final Agricultural Production, Consumption of Inputs, and Gross Value Added in the EC[1]

(Annual percentage changes)

	1973–82	1983	1984	1985	1986
Final production					
France	...	−2.6	3.1	1.3	0.4
Germany, Federal Republic of	2.0	−2.7	3.0	−3.4	4.1
Italy	...	6.5	−3.0	0.3	1.6
United Kingdom	1.7	−1.5	8.0	−3.5	−0.2
EC average	9.7	−0.3	3.0	−0.6	1.8
Consumption of inputs					
France	...	0.4	1.6	0.9	1.6
Germany, Federal Republic of	1.9	−0.2	−1.0	1.4	−1.8
Italy	...	1.6	0.2	0.5	1.5
United Kingdom	—	2.6	−1.7	−1.4	−0.1
EC average	7.3	1.2	0.1	1.2	0.5
Gross value added					
France	...	−4.6	4.3	1.5	−0.5
Germany, Federal Republic of	2.3	−5.3	7.2	−8.2	10.6
Italy	...	8.5	−4.1	0.3	1.6
United Kingdom	4.1	−6.1	19.7	−5.5	−0.4
EC average	11.8	−1.4	5.3	−1.7	2.6

Source: Commission of the European Communities, *The Agricultural Situation in the Community—1987 Report* (Brussels; Luxembourg, 1988).

[1] In real terms.

Table A50. Net Agricultural Value Added in the EC at Factor Cost per Manpower Unit[1]

(Annual percentage changes)

	1975–81	1982	1983	1984	1985
France	0.5	15.5	−4.8	−1.6	−2.4
Germany, Federal Republic of	−2.6	17.7	−20.6	16.2	−11.4
United Kingdom	−2.3	10.6	−8.3	17.9	−19.9
EC average	6.1	13.3	−13.2	6.5	−7.0

Source: Commission of the European Communities: *The Agricultural Situation in the Community—1987 Report* (Brussels; Luxembourg, 1988).

[1] In real terms.

Table A49. Employment by Sector in the EC

(In percent of total civilian employment)

	1960	1970	1980	1983	1984	1985	1986
Agriculture							
France	22.5	13.5	8.7	7.9	7.8	7.6	7.3
Germany, Federal Republic of	13.8	8.6	5.6	5.6	5.6	5.6	5.3
Italy	32.6	20.2	14.3	12.4	11.9	11.2	10.9
United Kingdom	4.8	3.2	2.6	2.7	2.6	2.6	2.6
EC average	18.4	11.4	8.0	7.6	7.4	7.2	7.0
Industry							
France	37.6	39.2	35.9	33.8	32.9	32.0	31.3
Germany, Federal Republic of	48.2	49.3	44.1	41.8	41.3	41.0	40.9
Italy	33.9	39.5	37.9	36.1	34.5	33.6	33.1
United Kingdom	47.6	44.8	37.8	33.7	32.9	32.4	31.1
EC average	41.5	42.6	38.0	35.3	34.5	34.0	33.5
Services							
France	39.9	47.2	55.4	58.3	59.4	60.4	61.4
Germany, Federal Republic of	38.0	42.1	50.3	52.6	53.1	53.4	53.8
Italy	33.5	40.3	47.8	51.5	53.6	55.2	56.0
United Kingdom	47.6	52.0	59.6	63.6	64.4	65.0	66.1
EC average	40.1	45.9	54.0	57.1	58.1	58.9	59.5

Source: Commission of the European Communities, *The Agricultural Situation in the Community—1987 Report* (Brussels; Luxembourg, 1988).

Table A51. Federal Republic of Germany: Coefficients of Nominal Protection for Selected Agricultural Products[1]

	1975–80	1981	1982	1983	1984	1985	1986
Beef	236.1	171.9	152.3	149.0	...	...	...
Sugar	163.1	111.6	221.6	211.0	298.6	363.3	331.0
Butter	274.0	175.9	171.1	176.8	174.4	197.6	262.2
Maize	196.6	169.7	194.8	155.6	133.6	145.4	227.5
Wheat	155.2	121.3	127.6	123.6	108.5	106.0	164.1

Sources: Statistical Office of the European Communities, *EUROSTAT, Agricultural Prices*; and International Monetary Fund, *International Financial Statistics*, various issues.

[1] Coefficients of nominal protection are defined as the ratio of domestic to world market prices expressed in percent.

Table A52. Federal Republic of Germany: Degrees of Self-Sufficiency in Selected Agricultural Products[1]

(In percent)

	1960–64	1965–69	1970–74	1975–79	1980–84	1985
Cereals	76.6	76.0	78.0	82.9	90.3	99.5
Wheat	75.9	83.5	83.7	97.6	104.3	...
Fruit and vegetables	60.9	55.7	47.9	40.9	44.9	47.3
White sugar	89.9	86.8	95.5	116.0	133.4	131.9
Beef	95.6	88.0	89.3	96.7	109.7	111.6
Pork	96.9	96.0	91.5	90.0	88.7	88.2
Butter	94.3	101.8	105.6	132.3	134.8	111.9
Eggs	65.4	84.3	84.1	77.5	71.9	72.5
Wine	59.5	56.8	62.7	56.8	59.5	55.5

Sources: Statistical Office of the European Communities, *Yearbook of Agricultural Statistics*, various issues; and Julius Rosenblatt and others, *The Common Agricultural Policy of the European Community: Principles and Consequences*, Occasional Paper, No. 62 (Washington: International Monetary Fund, December 1988).

[1] Defined as ratios of domestic production to consumption.

Table A53. EC Common Agricultural Policy: Budgetary Expenditures of the EAGGF-Guarantee Fund by Commodities[1]

(In billions of ECUs or in percent)

	1981 ECU	1981 Percent	1982 ECU	1982 Percent	1983 ECU	1983 Percent	1984 ECU	1984 Percent	1985 ECU	1985 Percent	1986 ECU	1986 Percent
Cereals and rice	2.0	17.8	1.9	15.1	2.5	16.0	1.7	9.3	2.4	12.0	3.5	15.8
Sugar	0.8	7.0	1.2	10.0	1.3	8.4	1.6	8.9	1.8	9.1	1.7	7.8
Fats and protein plants	1.1	9.9	1.3	10.5	1.8	11.1	2.0	10.7	2.2	11.0	3.1	14.0
Fruit and vegetables	0.6	5.8	0.9	7.4	1.2	7.6	1.5	7.9	1.2	6.2	1.0	4.5
Wine	0.5	4.2	0.6	4.6	0.7	4.2	1.2	6.7	0.9	4.7	0.6	2.9
Tobacco	0.4	3.3	0.6	5.0	0.7	4.2	0.8	4.3	0.9	4.4	0.8	3.5
Milk products	3.3	30.5	3.3	26.9	4.4	27.9	5.4	29.7	5.9	30.1	5.4	24.4
Meat, eggs, and poultry	1.9	17.1	1.6	13.2	2.3	14.6	3.3	17.7	3.5	17.6	4.3	19.7
Other markets	0.4	3.6	0.6	4.7	0.6	3.5	0.5	2.9	0.7	3.8	1.1	5.1
Agrimonetary measures	0.2	2.2	0.3	2.5	0.5	3.1	0.4	2.1	0.2	1.0	0.5	2.2
Other expenditures	−0.2	−1.5	—	—	−0.1	−0.7	—	−0.2	—	0.2	0.1	0.3
Total EAGGF-Guarantee Fund	11.0	100.0	12.4	100.0	15.8	100.0	18.3	100.0	19.7	100.0	22.1	100.0
Memorandum item EAGGF-Guarantee Fund (*in percent of total budgetary expenditures*)		61.6		60.6		64.9		66.6		70.2		64.7

Source: European Communities, *Official Journal of the European Communities*, C 336, Vol. 30 (December 15, 1987).

[1] European Agricultural Guidance and Guarantee Fund, Guarantee Section.

Table A54. Federal Republic of Germany: Payments to and Receipts from the EC

(In millions of ECUs)

	1980	1981	1982	1983	1984	1985	1986
Payments							
Customs duties	1,799.1	1,943.8	1,966.5	2,019.8	2,309.7	2,414.5	2,436.1
Agricultural levies	223.7	179.9	201.9	143.0	158.6	142.0	111.9
Sugar and isoglucose	130.7	126.9	190.2	270.5	350.2	286.3	313.0
Value-added tax[1]	2,456.7	2,806.5	3,339.9	4,038.8	4,233.9	4,661.5	5,869.2
Total own resources	4,610.2	5,057.1	5,698.5	6,472.1	7,052.4	7,504.3	8,730.2
(*In percent of EC total*)	29.9	28.1	26.9	28.1	28.4	28.8	26.2
Receipts							
EAGGF-Guarantee[2]	2,451.4	2,031.5	2,027.5	3,075.8	3,320.0	3,625.6	4,400.6
EAGGF-Guidance	142.1	134.1	107.1	107.7	89.3	81.0	105.1
Social Fund	80.5	72.3	89.9	81.5	63.8	109.8	134.6
Regional Fund	50.4	36.2	61.6	45.0	43.9	61.7	92.5
Fisheries	3.1	2.8	3.7	2.8	5.1	2.3	6.1
Specific measures	—	—	—	270.7	191.7	20.1	—
Reimbursement of costs incurred in collecting own resources	212.7	225.4	237.0	241.8	302.5	284.5	249.5
Total	2,940.2	2,502.3	2,526.8	3,825.3	4,019.3	4,185.0	4,988.4
(*In percent of EC total*)	20.1	16.1	14.0	17.7	16.7	17.0	16.4
Net contribution	**1,670.0**	**2,554.8**	**3,171.7**	**2,646.8**	**3,033.1**	**3,319.3**	**3,741.8**
(*Percentage change*)	...	53.0	24.1	−16.5	14.6	9.4	12.7

Source: European Communities, *Official Journal of the European Communities*, C 336, Vol. 30, (December 15, 1987).
[1] Including the balances and adjustments for previous financial years.
[2] EAGGF = European Agricultural Guidance and Guarantee Fund.

Table A55. Intervention Prices in the EC in 1987/88 for Member Countries[1]

(Percentage change from 1986/87)

	In ECU	In National Currency[2]
Germany, Federal Republic of	0.0	0.0
France	−0.2	4.2
Italy	−0.6	3.3
Netherlands	0.0	−0.5
Belgium	0.0	1.7
Luxembourg	0.0	1.6
United Kingdom	0.0	6.5
Ireland	0.0	8.6
Denmark	0.0	2.3
Greece	−0.4	13.3
EUR-10	−0.2	3.4
Spain[3]	1.4	7.2
Portugal[3]	0.3	8.6

Source: *European Community—News* (Washington), No. 17/87 (July 13, 1987), p. 6.
[1] Average weight according to the importance of the different products.
[2] Prices in national currency converted from ECU at green rates.
[3] Allowing for the alignment of Spanish and Portuguese prices to the CAP's common prices following EC enlargement.

Table A56. Federal Republic of Germany: Real Output and Input Prices in Agriculture

(In percentage changes)

	1983	1984	1985	1986	1987 (Estimate)
Output prices	−4.0	−3.5	−6.0	−5.4	−4.5
European average[1]	−2.6	−3.4	−4.3	−3.8	−4.5
Input prices	−2.5	−0.3	−4.0	−7.3	−5.1
European average[1]	−0.8	−0.6	−4.5	−6.3	−4.7
Ratio of output to input prices	−1.6	−3.2	−2.0	2.0	0.7
European average[1]	−1.9	−2.9	0.2	2.6	0.2

Source: Commission of the European Communities, *The Agricultural Situation in the Community—1987 Report* (Brussels; Luxembourg, 1988).
[1] European Community of 10.

Table A57. Rates of Capacity Utilization in the EC Steel Industry, 1970–83

(In percent)

	1970	1975	1980	1981	1982	1983
France	91.0	64.0	71.3	71.8	62.6	65.9
Germany, Federal Republic of	84.8	64.3	65.5	61.5	54.9	55.8
Italy	81.3	66.6	67.3	62.3	57.7	55.4
United Kingdom	...	74.2	40.3	61.1	55.2	61.1
EC	86.2[1]	66.1	63.1	63.4	56.2[2]	57.3[2]

Source: Ute Herdmann and Frank D. Weiss, "Wirkungen von Subventionen und Quoten − Das Beispiel der EG-Stahlindustrie," *Die Weltwirtschaft* (Tübingen), No. 1 (July 1985), p. 109.
[1] Excluding Ireland, Denmark, and the United Kingdom.
[2] Including Greece.

Table A58. Employment in the EC Steel Industry

(In thousands)

End of Period	1973	1980	1985	1986	July 1987
Belgium	62.4	45.2	34.5	30.5	28.4
Denmark	2.7	2.2	1.8	1.7	1.5
France	151.7	104.9	76.1	68.4	63.5
Germany, Federal Republic of	228.4	197.4	150.8	142.7	128.6
Ireland	0.7	0.7	0.5	0.6	0.6
Italy	89.7	99.6	67.4	65.6	64.2
Luxembourg	23.2	14.9	12.6	12.2	11.2
Netherlands	23.3	21.0	18.9	18.9	18.8
United Kingdom	196.2	112.1	59.1	55.9	53.5
EC (9) Total	778.3	598.0	421.7	396.5	370.3
Greece	...	...	4.2	4.1	4.1
Portugal	...	...	5.7	5.8	5.8
Spain	...	...	53.6	49.6	47.8
EC (12) Total	...	...	485.2	456.0	428.0

Source: Commission of the European Communities.

Table A59. Subsidies to the EC Steel Industry, 1980–85[1]

	Total Subsidies	Production of Rolled Steel	Subsidies Per Ton of Rolled Steel
	(In billions of deutsche mark)	(In millions of tons)	(In deutsche mark)
Belgium	10.3	45.7	225
France	21.5	97.7	220
Germany, Federal Republic of	9.9	175.3	56
Italy	28.6	116.7	245
United Kingdom	13.6	62.9	216
Other EC countries	3.5	...	...

Source: Ute Herdemann and Frank D. Weiss, "Wirkungen von Subventionen and Quoten − Das Beispiel des EC-Stahlindustrie," *Die Weltwirtschaft* (Tübingen) No. 1 (July 1985), p. 103.
[1] Total for the period; 1985 data are estimates.

Table A60. Finished Steel Products in the EC: Quotas and Steel Production[1]

(In thousands of tons)

	Quotas[2]			Production		
	1987	1988			1987	
	I–IV	I	II	1986	I–III	(In percent)
Hot-rolled coil	15,187	3,832	3,696	21,387	15,679	26.9
Cold-rolled sheet	13,025	3,160	3,220	13,497	9,698	16.7
Galvanized sheet[3]	—	—	—	3,522	2,894	5.0
Other coated sheet[4]	—	—	—	4,021	3,275	5.6
Quarto plate	4,837	1,346	1,296	4,750	3,636	6.2
Heavy sections	4,390	1,245	1,276	4,547	3,396	5.8
Wire rod[5]	8,992	—	—	10,905	8,116	13.9
Reinforcing bar[4]	—	—	—	7,431	5,568	9.6
Merchant bar[5]	7,156	—	—	7,984	5,936	10.2

Source: Commission of the European Communities.
[1] Excluding Spain and Portugal.
[2] Quotas are often revised upward during the quarter in response to unexpected demand. The table shows the initial quotas set at the beginning of each quarter.
[3] Quotas ended December 1986.
[4] Quotas ended December 1985.
[5] Quotas ended December 1987.

Table A61. Shipbuilding Capacity in the EC, 1976–86[1]

(In thousands of tons)

	1976	1980	1985	1986
Denmark	600	550	500	450
France	800	500	280	225
Germany, Federal Republic of	1,500	900	800	700
Greece	100[2]	100[2]	60	60
Italy	494	348	279	275
Netherlands	584	326	224	212
Portugal	110	130	90	85
Spain	1,000	1,000	445	445
United Kingdom	840	710	320	270
Total	6,028	4,564	2,998	2,722
Memorandum item				
Japan	10,770	7,030	7,030	7,030

Source: Organization for Economic Cooperation and Development.
[1] Excludes Belgium, whose capacity was estimated at 102,000 tons in 1986. Ireland currently has no shipbuilding capacity.
[2] Estimate.

Table A62. Employment in the EC Shipbuilding Industry, 1975–86[1]

	1975	1980	1983	1984	1985	1986
Belgium/Luxembourg	7,467	6,523	4,104	4,060	3,923	2,995
Denmark	16,630	11,400	11,200	10,300	10,200	7,000
France	32,500	22,200	21,000	16,940	15,058	13,730
Germany, Federal Republic of	46,839	24,784	25,966	22,189	22,260	18,184
Greece	2,316	2,672	2,812	2,000	2,000	2,000
Ireland	869	750	550	—	—	—
Italy	25,000	18,000	12,800	12,800	12,000[2]	11,570[2]
Netherlands	22,662	13,100	11,250	10,330	6,236	5,400[2]
United Kingdom	54,550	24,800	20,486	14,655	14,200	12,500
Total	208,833	124,229	110,168	93,274	85,877	73,379
Spain	...	...	...	...	18,000	18,000
Portugal	...	...	...	...	5,370	5,087

Source: Commission of the European Communities.
[1] Excludes ship repairing.
[2] Estimate.

Table A63. Real Stock Market Capitalization in Selected Industrial Countries

(In percent of GNP)

	1982	1983	1984	1985	1986	1988
Germany, Federal Republic of	10	14	14	24	25	16
Canada	35	37	35	43	45	47
United States	41	46	42	49	52	49
Japan	36	43	52	58	84	107
France	5	8	9	13	19	16
Italy	5	6	6	14	21	13
United Kingdom	41	48	59	64	79	86

Source: International Monetary Fund, *World Economic Outlook: A Survey by the Staff of the International Monetary Fund* (Washington, April 1988).

Note: The figures cover total market capitalization of all listed stocks at the end of the year, excluding stocks of foreign companies and investment funds. Total market capitalization is affected by the distribution of shareholdings between the corporate sector and households; the greater the proportion of shareholdings in the hands of the corporate sector, the greater is the stock market capitalization, for a given market judgment on the future profitability of companies. For an explanation of this point see McDonald (1988c). As a result, the relative importance of the stock market is likely to be overstated for countries where a relatively high proportion of equity is held by corporations. Corporate equity holdings are, for example, quite important in Germany (see Table A23).

Table A64. Federal Republic of Germany: Projections for 1990–91 Under Alternative Scenarios—Macroeconomic Variables[1]

(Annual average change in percent)

	Baseline	Alternative Scenarios		
		(1)	(2)	(3)
GNP	2.7	3.0	3.0	3.4
Domestic demand	2.9	3.4	3.3	3.8
Exports	4.5	4.2	4.8	4.6
Imports	5.4	5.5	6.1	6.3
Current account[2]	3.6	3.4	3.0	2.8
Employment	0.4	1.0	0.9	1.4
Rate of unemployment[2]	7.2	6.1	6.3	5.3
Consumption deflator	2.2	1.8	1.8	1.3

Source: Fund staff projections.
[1] In constant prices. Baseline = unchanged policies; scenario (1) = relative decline of agricultural/coal prices; scenario (2) = trade liberalization in iron, steel, and textiles and clothing; scenario (3) = (1) + (2).
[2] In 1991.

Table A65. Federal Republic of Germany: Projections for 1991 Under Alternative Scenarios—Selected Sectoral Variables

(In percent)

	Deviations from Baseline Under Alternative Scenarios		
	(1)	(2)	(3)
Output	−1.3	0.3	−1.0
Basic goods[1]			
Protected goods[2]	0.9	−1.9	−0.9
Traded goods[3]	0.8	0.7	1.5
Nontraded goods[4]	0.9	0.8	1.7
Employment	−3.0	0.6	−2.4
Basic goods			
Protected goods	1.1	−2.2	−1.1
Traded goods	1.2	1.1	2.3
Nontraded goods	1.4	1.2	2.7
Exports	−16.6	—	−16.4
Basic goods			
Protected goods	0.9	3.1	4.0
Traded goods	0.5	0.5	1.0
Imports	0.2	0.2	0.5
Basic goods			
Protected goods	0.1	12.4	12.5
Traded goods	0.3	0.1	0.4

[1] Agriculture and coal mining.
[2] Iron, steel, shipbuilding, and textiles and clothing.
[3] Other manufacturing and traded services industries.
[4] Mainly services.

Appendix II
A Dynamic Macro-Model

Table A66. Federal Republic of Germany: A Dynamic Macro-Model

Equation Identification	Equation	Number
I. Production		
Potential output	$\Delta lnQ_t = a * \Delta lnL_t + b * \Delta lnK_t + t$	(1)
Labor input	$L_t = g * TLF_t * (1 - NARU_t) * WH_t$	(2)
Total labor force	$TLF_t = WAP_t * PR_t$	(3)
Capital input	$K_t = d * Ipr_t + (1-r) * K_{t-1}$	(4)
Capacity utilization	$CU_t = GNP_t/PGNP_t$	(5)
Potential GNP	$PGNP_t = (1/v) * Q_t$	(6)
II. Demand		
GNP	$GNP_t = Cpr_t + Ipr_t + E_t + ST_t + FB_t$	(7)
Private disposable income	$YD_t = GNP_t - OTTR_t + IT_t$	(8)
Private consumption	$\Delta lnCpr_t = 0.75 * \Delta lnYD_t + 0.25 \Delta lnYD_{t-1} + c*(0.75 \Delta ln\, TOT + 0.25 \Delta ln\, TOT_{t-1})$	(9)
Private investment	$Ipr_t = w * (p * GNP_t - BSTR_t) + A_t$	(10)
Domestic demand	$DD_t = GNP_t + M_t - X_t$	(11)
III. Public sector		
Expenditures	$GGE_t = IT_t + E_t$	(12)
Interest and transfers	$IT_t = z * PGNP_t$	(13)
Revenue	$GGR_t = OTTR_t + BSTR_t + OR_t$	(14)
Business sector tax revenue	$BSTR_t = RSl_t * GNP_t$	(15)
Other tax revenue	$OTTR_t = RS2_t * GNP_t$	(16)
Other revenue	$OR_t = o * GNP_t$	(17)
Budget balance	$GB_t = GGR_t - GGE_t$	(18)
IV. External sector		
Exports	$\Delta lnX_t = al * \Delta lnWD_t - a_2 * (0.5_t\Delta lnREER_t + 0.3\Delta lnREER_{t-1} + 0.2\Delta lnREER_{t-2})$	(19)
Imports	$\Delta lnM_t = b1*\Delta lnGNP_t + b_2 * (0.5_t\Delta lnREER_t + 0.3\Delta lnREER_{t-1} + 0.2\Delta lnREER_{t-2})$	(20)
Foreign balance	$FB_t = X_t - M_t$	(21)

Table A66 (*concluded*). Federal Republic of Germany: A Dynamic Macro-Model

Equation Identification	Equation	Number
V. Prices		
Export prices	$\Delta ln PEX_t = 2/3 * \Delta ln PGNP_t + 1/3 * \Delta ln PIMP_t$	(22)
Import prices	$\Delta ln PIMP_t = \Delta ln PGNP_t - \Delta ln REER_t$	(23)
Domestic demand deflator	$DDFL_t = (GNP_t/DD_t) * PGNP_t - (X_t/DD_t) * PEX_t + (M_t/DD_t) * PIMP_t$	(24)
Terms of trade	$TOT_t = PGNP_t/PIMP_t$	(25)
VI. Nominal balances		
Current account	$CA_t = PEX_t * X_t - PIMP * M_t - TRA_t$	(26)
Government balance	$NGB_t = DDFL_t * GB_t$	(27)
VII. Employment		
Employment	$\Delta ln EMP_t = (1/a * (\Delta ln GNP_t - b * \Delta ln K_t - t) - \Delta ln WH_t$	(28)
Unemployment	$U_t = TLF_t - EMP_t$	(29)

Table A67. Federal Republic of Germany: Exogenous Variables of the Dynamic Macro-Model

NARU	=	Natural rate of unemployment
WH	=	Working hours
WAP	=	Working age population
PR	=	Participation rate
ST	=	Stockbuilding
E	=	Government expenditures for consumption and investment
RS1	=	Share of business sector tax revenue in GNP
RS2	=	Share of other tax revenue in GNP
WD	=	Partner countries' import demand
REER	=	Real effective exchange rate
PGNP	=	GNP deflator
TRA	=	External transfers
A	=	Autonomous investment

Appendix III
A Computable General Equilibrium Model

Table A68. Federal Republic of Germany: A CGE Model

I. Demand

(1) Demand for intermediate inputs:

(1)–(12) $INTERINPUT(DOM, i, j) = F1[Z(j), P(DOM, i), P(IMP, i)];$
$i = 1, 2, 3; j = 1, \ldots, 4$

(13)–(24) $INTERINPUT(IMP, i, j) = F2[Z(j), P(DOM, i), P(IMP, i)];$
$i = 1, 2, 3; j = 1, \ldots, 4$

(25)–(28) $INTERINPUT(DOM, 4, j) = F3[Z(j)]; j = 1, 2, 3, 4$

(2) Demand for productive factors:

(29)–(32) $LAB(j) = F4[Z(j), P(LAB), P(CAP), P(LAND)]; j = 1, \ldots, 4$

(33)–(36) $CAP(j) = F5[Z(j), P(LAB), P(CAP), P(LAND)]; j = 1, \ldots, 4$

(37) $LAND(j) = F6[Z(j), P(LAB), P(CAP), P(LAND)]; j = 1$

(3) Final demand

(a) Investment

(38)–(41) $INV(DOM, i) = F7[INV(TOTAL), P(DOM, i), P(IMP, i)]; i = 1, \ldots, 4$

(42)–(44) $INV(IMP, i) = F8[INV(TOTAL), P(DOM, i), P(IMP, i)]; i = 1, \ldots, 3$

(45) $INV(TOTAL) = NINV(TOTAL) / IPI$

(b) Household consumption

(46)–(49) $CONS(DOM, i) = F10[NCONS(TOTAL), P(DOM, 1), P(DOM, 2),$
$P(DOM, 3), P(DOM, 4), P(IMP, 1), P(IMP, 2), P(IMP, 3)]; i = 1, \ldots, 4$

(50)–(52) $CONS(IMP, i) = F11[NCONS(TOTAL), P(DOM, 1), P(DOM, 2),$
$P(DOM, 3), P(DOM, 4), P(IMP, 1), P(IMP, 2),$
$P(IMP, 3); i = 1, 2, 3$

(53) $CONS(TOTAL) = NCONS(TOTAL) / CPI$

(c) Government consumption

(54) $GOV(DOM, 4) = NGOV(DOM, 4) / P(DOM, 4)$

(d) Exports

(55)–(57) $EX(i) = F14[FCP(EX, i)]; i = 1, \ldots, 3$

II. Supply

(1) Imports

(58)–(60) $IMP(i) = F15[FCP(IMP, i)]; i = 1, \ldots, 3$

(2) Domestic supply

(61)–(64) $Z(i) = F16[P(DOM, j), P(IMP, w), P(LAB), P(CAP), P(LAND);$
$j = 1, \ldots, 4 \text{ but } j \neq i, w = 1, \ldots, 3]; i = 1, \ldots, 4$

Table A68. (*concluded*). Federal Republic of Germany: A CGE Model

III. Domestic Prices and World Market Prices

(65)–(67) $P(DOM, i) = FCP(EX, i) * XRATE * SUBSIDYRATIO(i) * 1/MARKUP(i);$
$i = 1, \ldots, 3$

(68)–(70) $P(IMP, i) = FCP(IMP, i) * XRATE * DUTYRATIO(i) * MARKUP(i);$
$i = 1, \ldots, 3$

IV. Market Clearing

(1) Markets for primary factors

(71) $LAB = \sum_j LAB(j); j = 1, \ldots, 4$

(72) $CAP = \sum_j CAP(j); j = 1, \ldots, 4$

(73) $LAND = LAND(1)$

(2) Domestic production

(74)–(77) $Z(i) = \sum_j INTERINPUT(DOM, i, j) + INV(DOM, i) +$
$CONS(DOM, i) + GOV(DOM, i) + EX(i); i = 1, \ldots, 4$

V. Miscellaneous Equations

(78) $GDP = CONS(TOTAL) + INV(TOTAL) + GOV(DOM, 4) + E - M$

(79) $E = \sum_i Ex(i); i = 1, \ldots, 3$

(80) $M = \sum_i IMP(i); i = 1, \ldots, 3$

(81) $Y = F81(EP/MP, GDP)$

(82) $EP = \sum_i w(x, i)FCP(EX; i); i = 1, \ldots, 3$

(83) $MP = \sum_i w(m, i)FCP(IMP, i); i = 1, \ldots, 3$

(84) $CPI = \sum_i w(cd, i)P(DOM, i) + \sum_j w(cm, j)P(IMP, j); i = 1, \ldots, 4; j = 1, \ldots, 3$

(85) $IPI = \sum_i w(id, i)P(DOM, i) + \sum_j w(im, j)P(IMP, j); i = 1, \ldots, 4; j = 1, \ldots, 3$

(86) $NCONS = F86(NABS)$

(87) $NINV = F87(NABS)$

(88) $NGOV = F88(NABS)$

(89) $TB = EP * E - MP * M$

(90) $P(LAB)R = P(LAB) / CPI$

Note: A more detailed exposition is given in Mayer (1988).

Table A69. **Federal Republic of Germany: Notation of the Variables of the CGE Model**

Variable name	Interpretation	Number
INTERINPUT(DOM, i, j)	Sector j's use of domestic intermediate inputs, delivered by sector i	16
INTERINPUT(IMP, i, j)	Sector j's use of imported intermediate inputs of type i	12
LAB(j)	Use of labor by sector j	4
CAP(j)	Use of fixed capital by sector j	4
LAND(j)	Use of land by sector j	1
Z(j)	Total production in sector j	4
INV(DOM, i)	Investment demand for domestically produced goods of type i	4
INV(IMP, i)	Investment demand for imported goods of type i	3
INV(TOTAL)	Total investment	1
CONS(DOM, i)	Consumption of domestically produced goods of type i	4
CONS(IMP, i)	Consumption of imported goods of type i	3
CONS(TOTAL)	Total consumption	1
GOV(DOM, 4)	Government consumption of domestically produced goods of type i	1
EX(i)	Exports of domestically produced goods of type i	3
IMP(i)	Imports of type i	3
P(DOM, i)	Price of domestically produced goods of type i	4
P(IMP, i)	Price of imports of type i	3
P(LAB)	Price of labor	1
P(CAP)	Price of capital	1
P(LAND)	Price of land	1
FCP(EX, i)	Foreign currency price of exports of type i in the world market	3
FCP(IMP, i)	Foreign currency price of imports of type i in the world market	3
XRATE	Exchange rate (measured in local currency per unit of foreign currency)	1
MARKUP (i)	Markup factor in foreign trade (covering transport, insurance, etc.)	3
SUBSIDYRATIO (i)	One plus the ad valorem rate of export protection	3
DUTYRATIO (i)	One plus the ad valorem rate of import protection	3
LAB	Supply of labor	1
CAP	Supply of capital	1
LAND	Supply of land	1
NINV (TOTAL)	Total nominal investment	1
IPI	Price index for investment goods	1
NCONS (TOTAL)	Total nominal consumption	1
CPI	Price index for consumer goods	1

Table A69 (concluded). Federal Republic of Germany: Notation of the Variables of the CGE Model

Variable name	Interpretation	Number
$NGOV(DOM, 4)$	Nominal government consumption	1
GDP	Gross domestic product	1
E	Total exports	1
M	Total imports	1
Y	Real income	1
EP	Export price index (in foreign currency)	1
MP	Import price index (in foreign currency)	1
$NABS$	Nominal absorption	1
TB	Real trade balance	1
$P(LAB)R$	Real wage	1
	Total	107

Table A70. Federal Republic of Germany: Selection and Values of the Exogenous Variables of the CGE Model[1]

(In percentage change from base period)

Variables	Base Line	Alternative Scenarios		
		(1)	(2)	(3)
$CAP(j)$ $j = 1, \ldots, 4$	0	0	0	0
$LAND(1)$	0	0	0	0
$MARKUP(i)$ $i = 1, \ldots, 3$	0	0	0	0
$SUBSIDYRATIO(1)$	0	(en)	0	(en)
$SUBSIDYRATIO(2)$	0	0	0	0
$SUBSIDYRATIO(3)$	0	0	0	0
$DUTYRATIO(1)$	0	(en)	0	(en)
$DUTYRATIO(2)$	0	0	−24	−24
$DUTYRATIO(3)$	0	0	0	0
$XRATE$	0	0	0	0
$NABS$	0	0	0	0
$P(LAB)R$	0	0	0	0
$P(DOM, 1)$	(en)	−4	(en)	−4
$P(IMP, 1)$	(en)	−4	(en)	−4
Memorandum Number of exogenous variables	17	17	17	17

[1] The notation (en) indicates that the variable is endogenous in this experiment.

Bibliography

Argy, Victor, "Choice of Intermediate Money Target in a Deregulated and Integrated Economy with Flexible Exchange Rates," *Staff Papers*, International Monetary Fund (Washington), Vol. 30 (December 1983), pp. 727–754.

Artus, Jacques R., "The Disequilibrium Real Wage Rate Hypothesis—An Empirical Evaluation," *Staff Papers*, International Monetary Fund (Washington), Vol. 31 (June 1984), pp. 249–302.

Bangemann, Martin, "Standort Bundesrepublik in Gefahr?" *Wirtschaftswoche*, No. 43 (October 16, 1987), p. 48.

Biedenkopf, Kurt H., "Zu Perspektiven der Wirtschaftspolitik" (mimeographed; Bonn, January 1988).

Blejer, Mario, and Ke-Young Chu, eds., *Measurement of Fiscal Impact—Methodological Issues*, Occasional Paper, No. 59 (Washington: International Monetary Fund, June 1988).

Board of Governors of the Federal Reserve System, *Federal Reserve Bulletin* (Washington, various issues).

Burda, Michael C., "Is There a Capital Shortage in Europe?" *Weltwirtschaftliches Archiv* (Kiel), Vol. 124 (1988), pp. 38–57.

―――, and Jeffrey D. Sachs, "Institutional Aspects of High Unemployment in the Federal Republic of Germany," NBER Working Paper, No. 2241 (Cambridge, Massachusetts: National Bureau of Economic Research, May 1987).

Burtless, Gary, "Jobless Pay and High European Unemployment," in *Barriers to European Growth: A Transatlantic View*, ed. by R.Z. Lawrence and C.L. Schultze (Washington: The Brookings Institution, 1987), pp. 105–162.

Coe, David T., "Nominal Wages, the NAIRU and Wage Flexibility," *OECD Economic Studies* (Paris), No. 5 (1985), pp. 87–126.

―――, and F. Cagliardi, "Nominal Wage Determination in Ten OECD Economies," OECD Working Paper, No. 19 (Paris: Organization for Economic Cooperation and Development, Economics and Statistics Department, March 1985).

Deutsche Bundesbank, *Monthly Report of the Deutsche Bundesbank* and *Supplements* (Frankfurt, various issues).

―――"The Bundesbank's Transactions in Securities Under Repurchase Agreements," *Monthly Report of the Deutsche Bundesbank* (Frankfurt), Vol. 35 (May 1983), pp. 23–30.

―――(1985a), "The Longer-Term Trend and Control of the Money Stock," *Monthly Report of the Deutsche Bundesbank* (Frankfurt), Vol. 37 (January 1985), pp. 13–26.

―――(1985b), "Freedom of Germany's Capital Transactions with Foreign Countries," *Monthly Report of the Deutsche Bundesbank* (Frankfurt), Vol. 37 (July 1985), pp. 13–23.

―――(1985c), "Recent Developments with Respect to the Bundesbank's Securities Repurchase Agreements," *Monthly Report of the Deutsche Bundesbank* (Frankfurt), Vol. 37 (October 1985), pp. 18–24.

―――(1987a), *The Deutsche Bundesbank: Its Monetary Policy Instruments and Functions*, Special Series No. 7 (Frankfurt: Deutsche Bundesbank, 2nd ed., October 1987).

―――(1987b), *Report of the Deutsche Bundesbank for the Year 1986* (Frankfurt: Deutsche Bundesbank, 1987).

―――(1988a), "Trends in the Euro-Deposits of Domestic Nonbanks," *Monthly Report of the Deutsche Bundesbank* (Frankfurt), Vol. 40 (January 1988), pp. 13–21.

―――(1988b), "The Economic Scene in the Federal Republic of Germany Around the Turn of 1987–88," *Monthly Report of the Deutsche Bundesbank* (Frankfurt), Vol. 40 (February 1988), pp. 5–42.

―――(1988c), "Methodological Notes on the Monetary Target Variable M3," *Monthly Report of the Deutsche Bundesbank* (Frankfurt), Vol. 40 (March 1988), pp. 18–21.

―――(1988d), "Recent Trends in Nonresidents' Investment Behavior in the Bond Market," *Monthly Report of the Deutsche Bundesbank* (Frankfurt), Vol. 40 (July 1988), pp. 13–18.

―――(1988e), "The Economic Scene in the Federal Republic of Germany in Summer 1988," *Monthly Report of the Deutsche Bundesbank* (Frankfurt), Vol. 40 (September 1988), pp. 5–42.

―――(1988f), *Report of the Deutsche Bundesbank for the Year 1987* (Frankfurt: Deutsche Bundesbank, 1988).

Donges, Juergen B., Klaus-Dieter Schmidt, and others, *Mehr Strukturwandel für Wachstum und Beschäftigung* (Tübingen: J.C.B. Mohr (Paul Siebeck), 1988).

Engels, W., "Tax Reforms in Germany," paper presented at the conference on "A Supply-Side Agenda for Germany?" held on June 29–30, 1988 in Cologne, Federal Republic of Germany.

Englander, Steven, and Axel Mittelstädt, "Total Factor Productivity: Macroeconomic and Structural Aspects of the Slowdown," *OECD Economic Studies* (Paris), No. 10 (Spring 1988), pp. 7–56.

European Communities, *Official Journal of the European Communities* (Brussels, various issues).

———, *The Agricultural Situation in the Community—1987 Report* (Brussels; Luxembourg, 1988).

———, EUROSTAT, *Agricultural Prices* (Luxembourg, various issues).

———, *Yearbook of Agricultural Statistics* (Luxembourg, various issues).

———, *European Community—News* (Washington; New York, various issues).

Federation of the German Stock Exchanges, *Annual Report, 1987* (Frankfurt: Federation of the German Stock Exchanges, 1988)

Germany, Federal Republic of, Bundesministerium der Finanzen, *Finanzbericht* (Bonn, various issues).

———, Bundesministerium der Finanzen, *Tasks and Objectives of a New Fiscal Policy: The Limits to Public Indebtedness* (Bonn, 1985).

———, Bundesministerium der Finanzen, *The Eleventh Subsidy Report* (Bonn, 1987).

———, Federal Government, *Jahreswirtschaftsbericht* [Annual Report] (Bonn, 1988).

———, Federal Government, *Aktuelle Beiträge zur Wirtschafts- und Finanzpolitik* (Bonn, various issues).

———, Ministry of Economics, *BMWI Tagesnachrichten* (Bonn, various issues).

———, Statistisches Bundesamt, *Aussenhandel* (Wiesbaden, various issues).

———, Statistiches Bundesamt, *Statistisches Jahrbuch* (Wiesbaden, various issues).

———, Statistiches Bundesamt, *Volkswirtschaftliche Gesamtrechnungen* (Wiesbaden, various issues).

———, Statistiches Bundesamt, *Wirtschaft und Statistik* (Wiesbaden, various issues).

Guitián, Manuel, "The European Monetary System: A Balance Between Rules and Discretion," Part I in *Policy Coordination in the European Monetary System*, Occasional Paper, No. 61 (Washington: International Monetary Fund, September 1988).

Gundlach, Erich, "Gibt es genügend Lohndifferenzierung in der Bundesrepublik Deutschland?" *Die Weltwirtschaft* (Tübingen), No. 1 (1986), pp. 74–78.

Heller, Peter S., Richard D. Haas, and Ahsan S. Mansur, *A Review of the Fiscal Impulse Measure*, Occasional Paper, No. 44 (Washington: International Monetary Fund, May 1986).

Herdmann, Ute, and Frank D. Weiss, "Wirkungen von Subventionen und Quoten—Das Beispiel der EG-Stahlindustrie," *Die Weltwirtschaft* (Tübingen), No. 1 (July 1985), pp. 101–113.

Institut der deutschen Wirtschaft, "Schwerpunkt: Standort Bundesrepublik Deutschland," *IW-Trends* (Cologne), Vol. 15 (February 1988).

International Monetary Fund, *International Financial Statistics* (Washington, various issues).

———, "European Community Issues Directive to Liberalize All Capital Movements," *IMF Survey* (Washington), Vol. 17 (August 29, 1988), pp. 283–84.

———, *World Economic Outlook: A Survey by the Staff of the International Monetary Fund* (Washington, various issues).

Jäckel, Peter, "Industriekonjunktur leicht aufwärts gerichtet," *IFO Schnelldienst* (Munich), No. 9 (March 1988), pp. 3–8.

Keesing, Donald B., and Martin Wolf, *Textile Quotas Against Developing Countries* (London: Trade Policy Research Center, 1980).

Kelly, Margaret, and others, *Issues and Developments in International Trade Policy*, Occasional Paper, No. 63 (Washington: International Monetary Fund, December 1988).

Kenward, L.R., "Perspectives on the International Competitive Position of the U.S. Steel Industry," IMF Working Paper, No. 87/8 (Washington: International Monetary Fund, February 1987).

Kirmani, Naheed, Pierluigi Molajoni, and Thomas Mayer, "Effects of Increased Market Access on Exports of Developing Countries," *Staff Papers*, International Monetary Fund (Washington), Vol. 31 (December 1984), pp. 661–84.

Kremers, J., "Aspects of Money Control in Germany" (unpublished; Washington: International Monetary Fund, 1988).

Layard, R., and G. Basevi, "Europe: The Case for Unsustainable Growth" (Brussels: Centre for European Policy Studies, May 1984).

Leibfritz, W., and R. Parsche, "Steuerbelastung der Werkzeugmaschinenindustrie im Internationalen Vergleich," *IFO Schnelldienst* (Munich), No. 9 (March 1988), pp. 9–16.

Lipschitz, L., "Wage Gaps, Employment, and Production in German Manufacturing" (unpublished; Washington: International Monetary Fund, 1986).

Madison, J., "No. 10: Madison," in *The Federalist Papers*, with an introduction by Clinton Rossiter (New York: New American Library, 1961), pp. 77–84.

Mayer, T., "Economic Structure, the Exchange Rate and Adjustment in the Federal Republic of Germany: A General Equilibrium Approach," IMF Working Paper, No. 88/79 (Washington: International Monetary Fund, September 1988).

McDonald, Donogh (1988a), "An Econometric Analysis of Machinery and Equipment Investment in Germany" (unpublished; Washington: International Monetary Fund, 1988).

———(1988b), "The Implications of Differential Growth Rates Among EMS Countries" (unpublished; Washington: International Monetary Fund, 1988).

———(1988c), "The Measurement of Stock Market Wealth and Its Impact on Consumption Behavior" (unpublished; Washington: International Monetary Fund, 1988).

Ochel, Wolfgang, and Manfred Wegner, "Dienstleistungen in Europa," *IFO Schnelldienst* (Munich), No. 14/15 (May 1987), pp. 11–15.

Olson, Mancur, *The Logic of Collective Action; Public Goods and the Theory of Groups* (Cambridge, Massachusetts: Harvard University Press, 1971).

Organization for Economic Cooperation and Development, *Development Cooperation* (Paris, various issues).

———, *Economic Outlook* (Paris, various issues).

———, *Main Economic Indicators* (Paris, various issues).

———, *Quarterly National Accounts* (Paris, various issues).

———, *Economic Surveys: Germany* (Paris, June 1986).

———, *The Tax/Benefit Position of Production Workers 1981–85* (Paris, 1986).

———, *Historical Statistics 1960–86* (Paris, 1987).

———, *Structural Adjustment and European Performance* (Paris, 1987).

———, *Economic Surveys: Germany* (Paris, June 1988).

Rosenblatt, J., and others, *The Common Agricultural Policy of the European Community: Principles and Consequences*, Occasional Paper, No. 62 (Washington: International Monetary Fund, November 1988).

Russo, Massimo, "Cooperation and Coordination in the EMS—The System at a Crossroad," in *The International Monetary System and Economic Development: A Challenge to the International Economic Cooperation*, proceedings from the Malente Symposium VII, ed. by Christian Dräger and Lothar Späth (Baden-Baden: Nomos Verlagsgesellschaft, 1988), pp. 281–312.

———, and Guiseppe Tullio, "Monetary Coordination Within the European Monetary System," Part II in *Policy Coordination in the European Monetary System*, Occasional Paper, No. 61 (Washington: International Monetary Fund, September 1988).

Sachverständigenrat zur Begutachtung der gesamtwirtschaftlichen Entwicklung [Council of Economic Advisors], *Vorrang für die Wachstumspolitik—Jahresgutachten 1987/ 88* (Stuttgart am Mainz: Verlag W. Kohlhammer, 1987).

Seffen, A., "Entwicklung der Krankenversicherungsfinanzen seit 1970," *IW-Trends* (Cologne), Vol. 14 (December 20, 1987), pp. D1–D10.

Soltwedel, Rudiger, and others, *Deregulierungspotentiale in der Bundesrepublik* (Tübingen: J.C.B. Mohr (Paul Siebeck), 1986).

Spinanger, Dean, and Joachim Zietz, "Managing Trade but Mangling the Consumer: Reflections on the EEC's and West Germany's Experience with the Multi-Fiber Arrangement," *Aussenwirtschaft* (Bern), Vol. 41 (1986), pp. 511–531.

Tanzi, Vito, "Tax Reform in Industrial Countries and the Impact of the U.S. Tax Reform Act of 1986," IMF Working Paper, No. 87/61 (Washington: International Monetary Fund, September 1987).

Thormählen, Thies, and Hans-Bodo Leibinger, "Sachverständigenrat: Zahlenmagier am Budget," *Wirtschaftsdienst*, No. 1 (1987), pp. 45–52.

Ungerer, Horst, and others, *The European Monetary System: Recent Developments*, Occasional Paper, No. 48 (Washington: International Monetary Fund, December 1986).

Weiss, Frank D., "Importrestriktionen der Bundesrepublik Deutschland," *Die Weltwirtschaft* (Tübingen), No. 1 (July 1985), pp. 88–100.

Witteler, D. "Tarifäre und nichttarifäre Handelshemmnisse in der Bundesrepublik Deutschland—Ausmass und Ursachen," *Die Weltwirtschaft* (Tübingen), No. 1 (1986), pp. 136–55.